AN INTRODUCTION

TO

GEOLOGY

AN INTRODUCTION

TO

GEOLOGY

BY

J. E. MARR, Sc.D., F.R.S.

PRESIDENT OF THE GEOLOGICAL SOCIETY
FELLOW OF ST JOHN'S COLLEGE, CAMBRIDGE

CAMBRIDGE :

at the University Press

1905

CAMBRIDGE
UNIVERSITY PRESS

University Printing House, Cambridge CB2 8BS, United Kingdom

Cambridge University Press is part of the University of Cambridge.

It furthers the University's mission by disseminating knowledge in the pursuit of education, learning and research at the highest international levels of excellence.

www.cambridge.org
Information on this title: www.cambridge.org/9781107426207

© Cambridge University Press 1905

First published 1905
First paperback edition 2014

A catalogue record for this publication is available from the British Library

ISBN 978-1-107-42620-7 Paperback

PREFACE.

IN the present volume an attempt is made to explain the scope and methods of Geology, without taxing the reader's memory with excess of detail, and avoiding so far as possible the use of technical terms.

It is hoped that it may be found useful to those general readers who wish to obtain some idea of the science but do not desire to pursue its study far, but especially as an introduction for those who will subsequently proceed to the perusal of more advanced treatises.

I am indebted to several friends for the use of photographs reproduced in the work. The figure of the pot-hole facing p. 39 is from a photograph by Mr Welch, of Belfast; the photograph of the glacier facing p. 51 was supplied by Prof. E. J. Garwood; that of the ice-worn rocks and perched stones by Prof. S. H. Reynolds; and that of the old Triassic landscape facing p. 174 was taken by Prof. H. E. Armstrong, F.R.S., and the print supplied by Prof. W. W. Watts, F.R.S. The photographs of the fossils and of the flint implements were kindly taken by W. G. Fearnsides, M.A., Fellow of Sidney Sussex College, Cambridge.

I have also to thank R. H. Rastall, B.A., of Christ's College, Cambridge, for reading through the proofs.

J. E. M.

CAMBRIDGE,
August 1905.

CONTENTS.

PLATES.

CHAPTER I.

INTRODUCTORY.

GEOLOGY is the science which has for its aim the knowledge of the past history of the earth. Much light is thrown upon this history by students of other sciences, but he who follows the pursuit of geology is above all beholden to the records which are stored in the rocks forming the crust of the earth, and in the present work we shall try to obtain from these rocks some insight into the story of our globe, leaving to others the question as to how that globe came into being.

We may get some idea of the methods by which the geologist pursues his studies, if we regard the things needful to the proper understanding of the history of mankind; for the studies of man and of the earth on which he dwells have much in common.

A book dealing with the history of England attempts to give an outline of the progress of the English people from the earliest times of which we possess any knowledge to the present day. No sudden break marks any period of that history, and the historian, in dividing his book into chapters, chooses the record of certain events as suitable for closing one chapter and beginning another; it is not needful that two historians writing of the same people should divide their books into chapters treating of the same periods, and it is certain that historians writing about separate nations would as a general rule not do this. So it is with the geologist.

But, whatever may be the nature of the chapter of a book recording the history of a nation, that history to be of value must be written so that events are recorded in the order in which they occurred. A history of our country in which the period of the Commonwealth was first taken into account, and subsequently that of the occupation of Britain by the Romans, then the reign of Victoria, and the Norman Conquest (in the order in which we have named them), would be useless. In the same way, a history of the earth in which the events were treated in their wrong order would be of no value. The geologist therefore, like the student of human history, must treat of the occurrences which it is his province to note and to interpret in definite and proper order. He must find some means of grouping the records of the rocks, according to their proper periods. We shall take into account the ways in which this is done in a future chapter. It is enough at this point to say that many of the rocks which form the earth's crust may be in a sense compared with volumes of books divisible into chapters, the chapters being further made up of a number of pages. It is true that the volumes of earth-lore are imperfect, the leaves being torn and crumpled, but as the result of much study we are able to piece together the information which they convey.

Now the student of human history might be able to divide the times of which it is his province to treat, into periods; he might furthermore be able to refer the various records which he possessed to their proper periods: but this would be of little avail if he had no knowledge of the conditions of existence at the present day. He could no doubt record the customs, the laws, the commercial transactions of former times, but of what use would this be if he were ignorant of the meaning of custom, of law, and of commerce? As it is, he is able to interpret the events of the past by his knowledge of the conditions and events of to-day; here again he is in exactly the same position as the geologist. If the latter were unaware of the present state of the earth and of the changes which are now taking place in it, he could not make out the story told by the rocks, even though able to arrange those rocks into groups in accordance with their respective ages.

The conditions of existence of man at the present day are to some extent known to anyone who begins to study human history, and it is therefore not needful for the beginner to study these conditions before reading the simple books which give a sketch of the history of our own nation. On the other hand, the conditions which exist upon the earth at present need study, and he who would get some insight into the science of geology, must know something of the present state of the earth.

This will be clear if we ponder for a moment upon the nature of the rocks of the earth's crust. One need not look at the rocks of a country for long before finding that, apart from their hardness, they are very like certain things which are being formed at the present day. The sandstones and mudstones which we so often meet with are very like the sands and muds which are seen along our sea-coasts; now, likeness of structure, if very close, usually means a similar mode of origin, and therefore; in order to find out the origin of the ancient sandstones and mud-

stones, we must study that of the modern sands and muds.
Again, we find other ancient rocks differing from sands
and muds, but similar to rocks which are formed by
existing volcanoes, and to know how these rocks were
made, we must know something about the volcanoes of
the present day.

There is another point wherein earth-history shews a
likeness to the history of man. In both, the records are,
on the whole, more obscure as we deal with periods more
remote from the present times. The student of English
history, for instance, when engaged with periods for the
study of which written records are available, finds that
the civilisation of those periods was not very widely
different from that of the present day, except in minor
points. When the people of Britain were in a state of
barbarism their mode of existence was however in many
respects different from that of ourselves, and it is not so
easy to restore the history of the period of which written
records are wanting, and the only relics left are weapons,
rude pottery, and the like. Still further back, when the
dwellers in these isles were in a state of savagery, even
relics begin to fail us, and we find that the dawn of
history is obscure, and such light as is thrown upon it
is due to study differing in nature from that pursued by
the ordinary historian.

The earth itself may also be regarded as having passed
through stages which we may compare with the savage,
barbaric and civilised stages of mankind. Of the first
stage the records of the rocks are silent, and we can only
infer what took place, on account of the light thrown
upon the story of the earth by other sciences, such as
astronomy. The records of the second stage are also
very obscure, and it will be well if the beginner pays

little heed to these, which need much study if we are to gain even a very little insight into their meaning. The third stage, which we may regard as having begun with the formation of the rocks which contain the earliest undoubted remains of once living animals and plants, is that which is as naturally the object of study of those who are but beginning geology, as the civilised period of man is of those who are taking their first lessons in human history.

CHAPTER II.

OF ROCKS.

As Geology is largely a study of the records of the rocks, it is clear that at the outset we must gain some knowledge of the nature of rocks. But as the nature of the rocks is in a high degree determined by the agents which produce them, and, on the other hand, the work of the agents is in turn affected by rocks upon which they operate, we are at once placed in a position of some difficulty. Luckily, however, the work of some of the more important agents is so readily seen and understood, that we shall not be seriously troubled if we begin with a short account of the rocks. It is true that in so doing we must refer now and again to matters which will be discussed more fully in later chapters, but only in the case of agents with which everyone already has some acquaintance.

What is a Rock? The student of a science often finds that a word is used in that science in a sense differing somewhat from that in which it is used in ordinary language, and this is the case with the word 'rock.' The usual idea of a rock connects the quality of *hardness* with the rock. He who follows the study of geology, however, will soon find that in many rocks other qualities are

more important than hardness. Indeed, the hardness
of a rock is often of no importance as bearing upon the
rock's origin. Special note is now made of this, because
we find that many of the ancient rocks are hard, and
many of those of modern date are soft, and therefore it
might be thought that there was a great difference between ancient and modern rocks. This is not so. In
certain cases, rocks which were formed quite recently are
very hard, while in other cases very ancient rocks are
found to be soft, or loose-grained.

To the geologist then, a mass of loose sand, such as
we find on the sea-shore or among the sandhills along
the coast, is as much a rock as a hard substance like
granite, and the student must at once grasp the fact of the
essential likeness of the rocks composed of loose particles
to others in which the particles are bound together.

A rock may be defined as a substance composed of
a number of mineral particles. It is true that this
definition is imperfect, for it requires a further statement
as to the nature of minerals, but for our present purpose
the popular idea of mineral substances will be enough,
if we merely state that, for the purposes of our definition,
particles of mineral matter which have once formed parts
of living things, such as bits of shell, are regarded as being
mineral particles.

Two great divisions of Rocks. A very brief inspection
of the earth's crust in suitable places will convince anyone
that there are many kinds of rocks. The chalk of the
southern cliffs of Albion, the grits of the Yorkshire moor-
lands, the granite of Dartmoor and the basalt of the
Giant's Causeway differ from each other in many respects.
But apart from minor varieties we may group all rocks
into two great classes, according to their mode of origin.

To one or other of these classes, all rocks, whether ancient or modern, can be referred. We may call these classes

(A) The Sedimentary Rocks.

(B) The Igneous Rocks.

A. *The Sedimentary Rocks.* If we take a basin of water and throw a few handfuls of loose sand into the water this sand will settle down as a *sediment* at the bottom of the basin. If we do the same at different intervals of time, taking an aquarium full of different kinds of creatures, we may find the remains of these creatures embedded in the sediment at the bottom of the aquarium. Now the ocean is like a great aquarium into which sand and other substances are carried by rivers or washed from the sea-cliffs by the waves. This material settles as sediment upon the bottom of the sea just as our handfuls of sand settled in the basin or aquarium. Similarly, we find sediments which have settled on the beds of rivers and lakes, or even upon the dry land. All of these sediments are formed of particles of pre-existing rocks, which have become re-arranged in a manner to be described more fully in a later chapter. They may and do vary greatly as regards their nature, but all possess this point in common.

B. *The Igneous Rocks.* Comparatively few people have seen an active volcano, but many of the readers of this book may have seen a blast-furnace, or at any rate some of the products which have come from a blast-furnace. The molten torrent of metal which issues from the furnace when it is tapped bears many resemblances to the *lava* which pours from the vent of a volcano, and very striking is the resemblance between the slags which accompany the metal, and some of the rocks which have issued from the volcano. Such rocks differ from sediments, inasmuch as they have solidified from a molten condition; they are therefore termed *igneous* rocks.

The differences between Igneous and Sedimentary Rocks. As these two classes of rocks differ as regards their mode of origin we should expect that they would possess differences of character, and this is to a large extent the case. The chief differences may be stated in a tabular form, and we may then discuss their nature.

Igneous rocks are:	Sedimentary rocks are:
(i) crystalline or glassy,	(i) usually not crystalline and never glassy,
(ii) nearly always without remains of organisms,	(ii) often marked by the inclusion of remains of organisms,
(iii) not arranged in layers called *strata*.	(iii) arranged in layers called *strata*.

(i) As a result of our knowledge of what occurs in glass-works and furnaces, we are aware that molten substances become solid in two very different states. In the one, the once molten material solidifies as a *glass*, the nature of which is so well known that no further remarks on it are required. In the other, the matter when solid occurs as a mass of *crystals*. These crystals have very well-marked properties, into the nature of which we cannot now enter, but many of these properties are constant for all crystals of any one mineral. They often possess the same shapes, the same degree of hardness, and the same lustre when light is reflected from their surfaces. We further know that the same matter may become solid as a glass or a crystalline mass according to the conditions under which it is solidified; thus, if a solid crystalline mass be melted and cooled very quickly it may assume the glassy state. Furthermore, we often

find that the degree of coarseness of a crystalline mass
may depend upon the rate of cooling; the slower the rate
of cooling the more coarsely crystalline the resulting solid
may be. We can then imitate the rocks which possess
the glassy structure, and to a certain degree those which
have the crystalline structure, by artificial means, and on
this account, and also from the similarity of many ancient
rocks to modern lavas which have been poured out from
volcanic vents, we are justified in concluding that these
ancient rocks have become solid from a state of fusion.
It must not be supposed, however, that the crystalline
structure is always readily detected. In many cases it
requires the aid of a microscope, but the student is recom-
mended to examine examples of crystalline rocks in which
the structure is readily seen, and to observe it carefully.
A piece of granite should be readily obtained and will
serve to illustrate the nature of the crystalline rocks.

The sediments, as has been seen, are composed of
particles of already existing rocks which have been broken
up and re-arranged. The broken surfaces of the particles
are frequently seen with ease, especially with the help of
a pocket-lens, and they are readily seen in many cases
to differ in outline from the crystalline particles of an
igneous rock, though there are other cases in which the
difference is not easy to make out.

Furthermore, these broken particles are often subjected to the rolling action of moving water, and become
ground against one another, whereby they acquire more
or less worn and rounded outlines, which help to enable
the student to recognise rocks composed of such particles.
The reader should obtain some sand from a sea-shore, or
river-bed (or from any available source), and make a careful
study of its component particles with the aid of a lens.

It will eventually be found that some sediments, for reasons into which we need not here enter, are composed of crystals. It is therefore obvious that the test of crystalline structure is in itself insufficient in all cases to separate an igneous rock from one of sedimentary origin.

(ii) It is very rare to find remains of organisms included in igneous rocks, for reasons which will be obvious. The reader is not likely to be troubled with such cases, and no further notice need now be taken of their existence. The sediments on the contrary are formed under conditions which render them extremely likely to become the burying places of the hard parts of living things such as fragments of wood, shells and bones. If we find these fragments of once living things embedded in rocks, the rocks are, in the highest degree of probability, sediments. Unfortunately the presence of these relics of life is far from being universal in the sediments, and therefore, although their presence indicates the sedimentary nature of the rock, their absence by no means points to its igneous origin.

(iii) The occurrence of sediments in definite layers is a matter of such prime import to the geologist that the discussion of the origin and nature of these layers requires a chapter to itself. It will be sufficient here to say that although different flows of lava piled up one above another may in some degree simulate the layers, beds, or *strata* (as they are termed) in which the sediments are arranged, it is merely a case of simulation. True strata are confined to the sediments, and the occurrence in strata is the most important of the characters which these sediments possess, as enabling us to read the events which go to make up the earth's history in the order of their occurrence.

CHAPTER III.

OF THE SEDIMENTARY ROCKS.

Stratification. If we took a flat-bottomed aquarium, and spread a layer of sand evenly over the bottom, and then spread a layer of mud evenly over the surface of the sand, there would be a flat plane between the sand and the mud. Similarly if a layer of sand be deposited naturally on a flat part of the sea-floor, and a layer of mud spread evenly over the sand, as in the case of the aquarium, a flat plane would separate the mud from the sand. Such a layer of sand or mud or of any other kind of material which is deposited is known as a *stratum* of rock, or more simply as a *bed* of rock; thus we talk of beds of sand, mud and limestone. The plane which separates one layer from another is known as a plane of *stratification*, or as a plane of *bedding*.

Suppose again, after a layer of sand had been deposited, there was a period during which no more sand was laid down. The grains of sand during this period might become in some degree attached to one another so that the rock was no longer a mass of loose grains. If after this period of pause, another layer of sand was laid down, as the first layer was now in some degree solid, the second layer would be separated from the first by a plane surface, which in this case divides two beds of similar character. This plane is also a bedding plane. It is clear therefore that bedding planes may separate two like or two unlike beds from each other. As, in many cases, the distance over which beds are laid down are very great as compared with the thickness of a single bed, the top and bottom of a bed are usually parallel. If then we look at a cutting, as for instance in a quarry, where a number of flat beds are shewn in an upright cliff, and where, therefore, the bedding planes appear on the surface of the cliff as straight lines, the beds will appear somewhat as in Fig. 1.

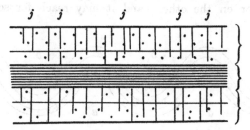

Fig. 1. Horizontal Beds in a quarry. *j j* Joints.

When a great number of bedding planes occur in a very small thickness of rock, say ten or fifteen to an inch, the layers are called *laminæ*, and the plane which separates two such layers is known as a plane of *lamination*. There is of course no real difference between a plane of lamination and a plane of stratification. A rock which is laminated is known as a *shale*. The central bed in Fig. 1 is a bed of shale divided into many laminæ.

The floor of the ocean has been found, as the result of many soundings, to be very flat, except in places near the shore. The result of this flatness is that beds as a rule are also flat when just formed. There are of course exceptions, but even then the slope of the beds is usually not at a very large angle.

Now we very often come across beds among the rocks of our land surfaces, where the bedding planes slope down towards the earth's interior at high angles, sometimes they are even upright or actually turned over. It is clear that the position of these beds has been shifted since they were formed, and indeed we often find that these sloping beds form portions of great arches and troughs as shewn in Fig. 2. These arches and troughs are of various sizes; one arch or trough may be only a few feet across (or even less), or on the other hand it may reach for scores of miles.

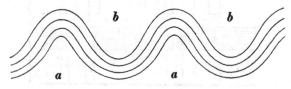

Fig. 2. *a a* arches, *b b* troughs.

We shall consider how the beds were thrown into these folds in a future chapter; at present we must consider some of the terms which are used in connexion with folded beds.

Suppose that the slates on one side of a house-roof be taken to represent a sloping bed of rock. We speak of a sloping bed as *dipping*, and we require to know (i) the direction, (ii) the amount, of the dip.

As regards direction, a bed dips in the direction in which it slopes *downwards* into the earth, so that if the house-roof sloped to the south, we should speak of the slates thereof, if they were really a bed of rock, as dipping to the south. But, you will say, there is a slope in every direction except along the level. That is so, and here we must consider the amount of the dip. There is one direction along the slates of the roof where the slope is steepest. If we were to pour water from a jug on to the roof, and the roof were even, the water would take that direction. That direction is the direction of *true* dip, but we usually speak of it simply as the *direction of dip*. The steeper the slope of the roof, the *higher* is the dip, which is usually recorded according to its angle. Thus if the house-roof were inclined at an angle of 30 degrees to a level plane, it would be said to dip at 30°, if at an angle of 45 degrees to the same plane, the dip would be 45°, and so on.

Now, if you will examine the level top of the roof, you will find that it extends in a direction at right angles to that of the dip. This level line at right angles to the direction of dip is spoken of as the direction of *strike* of a bed: thus if a bed dips to the north, its strike is east and west.

As the edges of beds come to the surface of the earth, it is clear that portions of the beds have been removed. This removal is brought about in a way which we shall soon regard. When the beds have been folded into arches and troughs, the tops of the arches have been swept off as shewn in Fig. 3, which shews some folded beds as they would appear if we made a deep upright cutting in the earth's crust. The level line shews the earth's surface at the top of the cutting, and the dotted part above this line shews those parts of the beds which have been removed. The student is at this point of his reading advised to fold up a number of stiff sheets of cardboard (which may be taken to represent beds) in the manner shewn in Fig. 3, and to cut off the tops of the arches along a level plane with a sharp knife, and then to consider the dip and strike which these pieces of cardboard would possess if they were in very truth beds of rock.

The reader, if he does as advised, will see that after the tops of the arches are cut off along a level plane, each piece of cardboard appears on the flat surface as a strip,

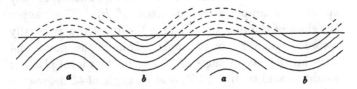

Fig. 3. Section through beds folded into arches, *a a* and troughs, *b b* with the tops of the arches removed by erosion.

and the strips of the different pieces of cardboard run parallel to one another. The place where beds come to the surface of the ground in the manner seen on looking at the cardboard strips is spoken of as the place where the beds *crop out*; each strip of a bed on the surface of the ground forms the bed's *outcrop*. Now on flat ground the direction of the outcrop is that of the strike of the bed, that is, it is at right angles to the direction of dip. On uneven ground the direction of outcrop does not always agree with that of the strike; but this is a subject which we will not here pursue.

Let us now regard for a moment the importance of finding out the dip and strike of beds by two examples. Suppose you knew that a valuable coal-seam cropped out two hundred yards west of some property which you owned, and that it dipped eastward. It would therefore lie below your ground at a depth below the surface depending on the angle of dip. If the dip was at a small angle, say ten degrees, the seam would not be so far below the surface of your property as if the angle of dip was higher, say sixty degrees. In one case the cost of boring a shaft to reach the coal might not be too great; in the other case the cost might be so great that it would not pay to bore for the coal.

Again, suppose you want to follow the coal along the surface of a level tract of ground. It may not always be seen at the surface; it may for instance be covered up beneath the soil. If you know the direction of dip, you also know that of the strike, which as we have seen is always at right angles to that of the dip. But as the line of outcrop is in the same direction as that of the strike on level ground, we should, knowing the strike, know where to seek for the outcrop of the coal below the soil at some distance from the point where the coal was first seen.

These are only two examples of the value of knowing rightly the direction of dip and strike of beds. That value is so great to the geologist that the student is most earnestly urged to pay great regard to the above account, for it will be needful to refer to dip and strike of rocks again and again as we proceed to consider other parts of our subject.

It will be convenient at this point to refer to a few terms used in connexion with folded rocks, which must be used in subsequent chapters.

When beds dip away on either side of a line or axis, the fold may be spoken of as a *saddle*. When they dip towards such an axis, the fold is a *trough*. In Figs. 2 and 3 *a a* shew saddles and *b b* troughs, as they would appear in an upright cutting. If the beds dip away on all sides from a central point instead of from a line, the fold is called a *dome*, and if they dip towards such a point, it is a *basin*. Beds which are thrown into a number of saddles and troughs are called *contorted* beds. The beds represented in Figs. 2 and 3 are therefore contorted.

CHAPTER IV.

OF DIVISIONAL PLANES IN ROCKS.

In addition to the planes of bedding, which were produced during the deposition of rocks, other planes are found affecting them, which require some notice.

Joints. If you enter a quarry of sandstone or granite you will probably notice a number of fissures in the rock, which in many cases are nearly or quite upright. In the sandstone quarry it may be observed that the most important of these lines run at right angles to the bedding planes, or nearly so, and they are obviously not bedding planes. They are in fact cracks in the rock, and are spoken of as *joints*. In the case of the sediments it is often found that one important set of joints traverses the rocks in the direction of their dip and another in that of their strike. These two systems are called *dip joints* and *strike joints*. Joints may be caused by more than one process. Some are due to shrinkage owing to loss of water or in the case of igneous rocks to consolidation, for the solid rock occupies less space than the molten rock. Others again may be formed by earth movement. The rocks when bent may eventually snap across, giving rise to joints. In sedimentary rocks, the important joints which follow the direction of dip and strike are *quadrangular* joints. The most regular of the joints produced

during the cooling of igneous rocks often assume the
form of six-sided columns, as seen in the basalt of Staffa
and the Giant's Causeway.   These are *columnar* joints.

In Fig. 1 the upright lines cutting the beds of sandstone represent joints.

Faults.   When beds are displaced along a joint, so
that one end of a stratum abuts against a different
stratum on the opposite side of the crack, the beds are
said to be *faulted*.   A *fault* is represented as it would
appear in an upright cutting in Fig. 4.   In this figure the

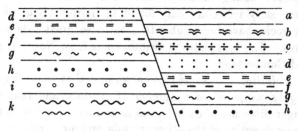

Fig. 4.   Section across a Fault.

right-hand side, where any bed, as *d*, appears at a lower
level than that of the corresponding part of the bed on
the other side of the fault, is known as the *downthrow*
side, while the other is the *upthrow* side.   The fissure
may be upright or inclined.   In the latter case it usually,
though by no means always, approaches the upright position.   Faults vary in the amount of displacement or *throw*.
The beds displaced may be shifted a fraction of an inch,
or on the other hand to the extent of thousands of feet.
Faults are produced by movements of the earth's crust.

Cleavage.   If you will examine a fairly thick roofing
slate, and put a chisel into one of its edges, and apply a
hammer, you will probably find that it may easily be split

into thinner slabs with faces parallel to those which form
the surfaces of the slate. That these planes along which
the slate thus splits are not bedding planes is clear,
insomuch as you can often see the beds running across
the face of the slate in a direction quite different from
that of the other planes. These planes along which a
rock cleaves are known as *cleavage planes*, and all rocks
which have this property of cleavage are known to the
geologist as *slates*. It is generally admitted that this
cleavage is impressed upon slates as the result of great
pressure acting sideways upon the rocks. In Britain
these slaty rocks are only found in limited areas, where
the rocks have been submitted to these great side pres-
sures. The best slates of Britain come from North Wales
and Cumberland.

Foliation. In some regions we find that the rocks are
largely composed of a number of crystals, and that the
longer direction of each crystal is arranged parallel to the
longer direction of the other crystals, and furthermore that
this direction coincides with that of certain divisional
planes, which to some extent recall those of lamination,
though they are often much more puckered than the
planes of lamination. These crystalline rocks possessing
such parallel structures are called *schistose* rocks or *schists*.
They are largely found in the Highlands of Scotland, and
abroad we find them abundant in parts of the Alps,
Norway, and other places. The origin of these schists is
still an obscure question, and it need not be entered upon
here.

To sum up, rocks may be affected by (i) planes of
bedding, (ii) joints, which when accompanied by displacement of the rocks form faults, (iii) cleavage planes, (iv)
foliation planes.

CHAPTER V.

OF THE PRINCIPAL GEOLOGICAL AGENTS.

HAVING got some idea of the characters of the rocks which form the crust of the earth, we are now in a position to discuss the effects of different agents which bring about changes in these rocks.

In former days, those who thought at all about the origin of the earth and the changes which it has undergone since it was formed believed that these changes were caused by agents of a very different nature from those which are at present in operation. It has for some time been admitted however that the changes which took place in past times, the records of which can be observed by inspection of the rocks of the earth's crust, were due to agents very similar both in kind and in degree to those which are at present working upon the earth. It is obvious, therefore, that in order to interpret the changes of the past, we must get some idea of the events of the present.

It will be convenient if we group the agents, other than those affecting life, under three heads, viz. :—

   (i)   Those at work upon the surface of the land.

   (ii)   Those at work in the sea.

   (iii)   Those operating under the earth's surface.

Furthermore, as those which are placed in the first division are most easily studied, being directly within our ken, we will begin by discussing the work of those agents which operate upon the surface of the land.

Before doing this, however, it may be well to give the reader a general idea of the class of work which is carried on by each group of agents.

(i) *Agents at work upon the surface of the land.* Variations in the temperature of the atmosphere, fall of rain and flow of rivers, all unite in performing a particular kind of work, namely the breaking up of solid rocks into loose fragments, and the carriage of these fragments from one place to another; and as the fragments on the whole travel downward rather than upward, the loosened portions of rock are by degrees carried toward and ultimately into the ocean. Ages may lapse before the bit of rock which is broken from the mountain peak finds its way to the sea, but should no other agent prevent its travel, the sea is its ultimate destination, though it may of course be reduced to many smaller fragments before it reaches its goal.

We shall presently consider in some detail the various processes which occur during this loosening of the rocks and their downward transport, but in the meantime we may give a short account of what, in the main, does happen.

In frosty weather a wanderer among the mountains will often hear a fragment of rock falling from some cliff on to the slope beneath. It has been detached by the frost and has started its journey downward. In spring, during the melting of the snows on the upland heights, the hillsides are often seamed with torrents hurrying down to join the rivers in the valleys beneath. Our fragment, with many another, may be washed down by one of these torrents and carried into the main stream of the valley. There it is knocked against other fragments and perchance broken into smaller pieces, and each of these may be hurried downward during the seasons of flood, until at last they are carried into the water of the ocean at the river's mouth.

In regions of 'eternal frost,' as the high Alps or the icy wastes of Greenland, the material which has been broken from the solid rocks is also carried away from its source, but here the agent of transport is no longer the running stream, but the slowly creeping glacier or mass of ice.

In the arid desert where stream action is often slight or in many parts wanting, the loosened fragments are carried onward by yet another agent, the wind which sweeps over the desert waste.

These are samples of the processes of loosening and carriage of rocks. As the result of these and similar processes, the surface of our land masses is gradually being lowered, and the wreckage of the land carried toward the sea. If no other change were to intervene, and if a sufficient lapse of time be granted, our continent and island masses would be at last reduced to sea-level, and the world would be occupied by a universal ocean. In a word, the effect of the agents operating upon the surface of the land is the *destruction* of that land.

(ii) *Agents at work in the sea.* The waves of the sea shore may be readily observed to assist those agents which are working over the general surface of the land in the destruction of that land. The storm-waves hurl the pebbles and sand of the coast-line against the cliffs, and these cliffs are gradually worn away.

But this work of the sea is confined to the coast-lines. Away from the coasts the sea serves as a receptacle for the material which has been carried from the land by the processes which we have briefly considered. As the rivers enter the sea the current is checked, and the rock-grains which have been borne onwards by the river current are compelled to drop. By the aid of sea-waves, tides and currents they may be carried for some distance from the shore, but they must at last find a resting-place on the floor of the ocean, there to form new *deposits*.

We must here notice another action of the rivers. Not all of the material which they carry seawards is visible. If we take a glass of water from a river in flood it may be turbid, and if we allow the water to stand, it gradually becomes clear, for the mud which caused its turbidity settles at the bottom of the glass. If we now strain off the water and boil it, we may perhaps find other material which existed in that water. This is specially noticeable in the water of rivers in a chalk district, and as a result of the presence of this material, the well-known 'fur' as it is termed is deposited in kettles. Chalk is formed of a material known as carbonate of lime, and in some conditions this material may be dissolved by water; in fact it melts in water, just as a piece of sugar does.

Now when this limy material reaches the ocean, it cannot be reduced to the solid condition by boiling or heating the ocean, for the ocean contains far too much water. But myriads of animals and even some plants are capable of extracting this limy material from sea-water and restoring it to the solid state to form parts of their substance. Thus the ordinary shells of the shellfish which live in the sea have been obtained from dissolved lime, and, accordingly, in addition to the deposits formed of material washed down in the visible state by rivers, we may and do find *organic deposits* formed of the hard parts of various living organisms.

The work of the sea then is of prime import to the geologist, for it is to the action of the sea that we owe the great bulk of those deposits which really constitute the volumes in which the earth's history is recorded.

The work of land agents being one of *destruction*, that of marine agents is, on the contrary, in the main one of *construction*.

(iii) *Agents operating under the earth's surface.* It has been seen that as the result of the agents operating on the surface of the land, the tendency is to destroy those lands, granted a sufficient lapse of time; and there is no doubt that in the course of geological ages there has been time to destroy our continents again and again. Their actual existence therefore implies that some other agents have been at work in addition to those which we have noticed. We get some idea of what these are if we visit a district possessing active volcanoes. Not only do we find rock brought out from below in a molten or a fragmental condition through the throat of a volcano, but as an accompaniment of volcanic eruptions the earth's crust is often violently shaken. These shakings or earthquakes are, however, by no means confined to volcanic regions; many parts of the earth are affected by them, and they sometimes occur even in our own country. It has been actually observed that the level of the land is often altered as the result of earthquakes, and in this way tracts of country may be uplifted, as was observed to be the case, for example, during a great earthquake which occurred in New Zealand in 1855. But though the movement of the land during earthquakes is noticeable, we have every reason to suppose that the movement often occurs so slowly that it gives us no direct evidence of its occurrence; and it is probable that this very gradual movement causes much more important change of level than the occasional jars which affect the crust during a period of earthquake. In these earth movements then we have a means of restoring our continents which would otherwise be completely destroyed. By them deposits formed below the sea may be brought above the level of the ocean surface, and it is to them that we owe the contortions of the rocks which have been described in a preceding chapter. The causes of these movements are little understood, but it is known that they are due to changes which take place beneath the surface of the earth.

As the result of this alteration of the relative level of land and sea, continents may be reproduced.

To sum up, so far as our land masses are concerned, in general the agents which operate on the land surfaces are agents of *destruction*, those which work in the sea, of *construction*, and those which work underground, of *reproduction*.

CHAPTER VI.

OF THE AGENTS AT WORK UPON THE SURFACE OF THE LAND.

IT has already been seen that, in order to make new deposits from rocks which are already in existence, two processes are needful. The first process is the breaking up of the solid rocks, and the second is the carriage or *transportation* of the broken materials. During this carriage, further fracture of existing rocks may take place, as will be noticed later.

The loose materials which are at the outset of the process of destruction supplied from the rocks of the earth's crust are loosened owing to the action of the weather, and accordingly the acts of loosening are spoken of as *weathering*.

It will be well if we consider the work of destruction of rock under three heads, viz.:—(i) *Weathering*, (ii) *Transportation* of the weathered materials, (iii) Further breakage of rocks during the process of transportation.

Weathering. The principal agents which are concerned in the breaking up of solid rocks during the acts of weathering are (i) changes of temperature of the air, (ii) freezing of water, and (iii) rain ; the processes are often aided by (iv) the presence of various forms of life.

(i)　It is well known that a large number of substances expand when heated and contract when cooled. A bar of iron for example is somewhat longer when hot than it is when cold. If the reader will examine the lines of a railway he will find that a space is left between two adjoining rails This is to allow for expansion in summer. If the rails were laid lengthways without these gaps, the lines would crumple during hot periods. Similarly the various particles of rocks lengthen when heated and shorten when cooled, and when the rocks are solid and no spaces are left between the various particles, the rock will be broken up as the result of lengthening of the particles by day and in summer, and their shortening by night and in winter. In England the effects of this are not very noticeable, as other agents play a more important part in the breaking up of rock masses, but in deserts the effects of the great changes of temperature which occur in those regions are very marked, and fragments of rock are frequently broken from the parent masses.

(ii)　The effects of water freezing in the crevices of rocks is very marked in our own country, and is still more pronounced in alpine and arctic regions. Water yields an exception to the general rule that substances occupy more space when hot than they do when cold. When water is at a temperature near its freezing point, a further fall of temperature causes it to expand, and this expansion becomes very marked when the water passes into the state of ice. The well-known tendency of water-pipes to burst during a frost is due to this cause. We may regard all the cracks and crevices in rocks as so many natural water-pipes, and when these are filled with water, and that water becomes frozen, the consequent expansion may cause the fracture of the rocks. We often see the roads covered with crumbled material after frost, owing to this process, the hardened surface of the road being broken up by the freezing of water between the particles which compose it, and large masses of rock may be detached from their parent-rock by the water which occupies the joints and other fissures of the rock becoming frozen. The slopes of rock-waste which are frequently found at the bases of precipices in temperate and arctic regions are largely supplied with material which has been broken from the cliffs above by the agency of frost.

(iii) The effects of rain are rather more complex in their nature than those due to changes of temperature and to the action of frost.

The mere beating of the rain-drops against rocks produces little effect, unless the rocks are very soft. Clay may be to some extent broken up in this manner, but most rocks are sufficiently hard to withstand the battering action of the rain.

It is the power which rain has of dissolving certain of the materials which exist in some rocks which causes rain to exert an important effect in breaking up those rocks. Some materials can be dissolved in pure rain-water, but rain often contains other substances besides water, and it is these which give rain its principal destructive power. Some of the gases which exist in the air render rain thus potent, and we must make special mention of a gas which when taken up by the water gives rise to what is known as *carbonic acid*. The action of acid in corroding many substances is well known, and this carbonic acid is able to dissolve many materials which occur in rocks, and accordingly when rain is charged with carbonic acid, it often frets the surface rocks of the earth's crust. Many rocks are often seen to be covered with a brown crust, which is

due to the action of acidulated water upon the rocks, and
the brown crust in many cases is of a crumbly character
very different from that of the hard rock from which it
has been derived. The very existence of materials in
solution in river·water shews that those materials have
been dissolved from rocks, for the rain which furnishes
the river-water does not contain such substances when it
falls upon the surface of the ground, and it has therefore
taken them up after it has reached the earth.

(iv) The action of living creatures upon rock weathering is also varied. These creatures help the processes of
rock destruction by wedging asunder rock masses, and
also by supplying materials which act as solvents, thus
aiding the·solvents which have been obtained by rain-
water from the air. One frequently sees the roots of
trees breaking down walls. The root gets into a crack in
the wall, and as this root grows it forces the portions of
the wall asunder and may eventually cause them to fall.
Similarly tree roots break fragments of rock from the
parent mass when they have forced their way along joints
and other planes of weakness. But what is true of tree
roots is also true of the roots of other and smaller plants.
The action is the same though on a different scale.

Burrowing animals as rabbits, badgers, and even worms
may and do break up rocks to some extent. It is true
that they do not produce appreciable effects on what we
are accustomed to style the solid rocks, but they do loosen
such rocks as hardened clays and sands.

When plants decay, they supply carbonic and other
solvent acids to rain-water, and aid in the work of solution.
We often find great masses of moss attached to rock
surfaces and when we tear the moss away we find a hollow
underneath having the exact outline of the boundary of

the moss. Of course, in some cases the hollow was there before the moss, but we have evidence that in other cases the presence of the moss has caused the hollow, partly because it supplies solvent substances, and partly because it acts like a sponge and holds the water against the rock for a longer time than it would remain there, if the moss were absent.

Transportation. There are three important kinds of agents of transportation which work upon the land, namely, wind, rivers, and masses of moving land-ice. As the action of rivers is most marked in effecting transportation of material in our own country, we will notice this first.

Transportation by rivers. If a sheet of water be still, it cannot carry materials. If it be moving, it is, owing to its motion, capable of performing a certain amount of work, which, other things being equal, is greater as the rate of motion becomes greater. Accordingly rivers can carry more material as their swiftness increases, and swift rivers can also move heavier rock fragments than slow ones can carry.

We have already noted that the material which is carried down by streams is partly in *solution*, and partly in *suspension*. At present we will confine our attention to the material which is carried in suspension. This is borne onward in two ways. Some of the material, and this the coarser, is swept along the floor of the river; the other and finer material is actually held up in the water, and may be distributed throughout the water from the river's bottom to its surface. When the river is carrying pebbles, these will most probably be pushed along the floor, while the sand and mud will be more prone to be dispersed through the whole depth of the stream.

The material which is dissolved in the water of rivers varies considerably in character and in amount according to the nature of the rocks over which the water flows. The most important substances dissolved in river-water are compounds of lime.

A few words may be added as to the mode of introduction into the river-water of the matter which is transported by it.

The rock-waste which has been derived from the solid rocks by processes of weathering lies over those rocks in places, becoming more and more broken up as it is exposed to the action of the weather, until it forms a kind of blanket of loose particles covering the hard rocks beneath. During heavy rains, the rain-drops collect in hollows and form little runnels which course down towards the river-beds, and as each runnel may be charged with some of the loose rock-waste, a considerable quantity of this waste is carried into rivers during a period of rainfall, and it is this which forms so large a proportion of the material which rivers transport.

Transportation by wind. A certain amount of material is carried by wind even in our own country. The transportation is affected by several things, including the slope of the country, for it is easier for material to be blown downward than upward; the direction of the prevailing wind, for the material will on the whole be blown away from the direction from which the wind blows; and the state of the material over which the wind blows. In our own country, except along the sandhills of the coasts, so much of the rock-waste is bound together by the tangle of vegetable matter which grows on it, that it is able to a great extent to resist the action of the wind. It is in deserts, where the vegetation is scanty, that the effects of

wind as an agent of transportation are most marked.
There the fierce wind-blasts whirl the particles of rock-
waste high in the air, and often bear them onward for
great distances. When comparing the transporting action
of wind with that of rivers, we notice that in each case
some of the material is swept along the surface of the
earth, while the lighter particles are carried at various
heights above that surface. A noticeable difference is
that whereas in the case of rivers the transport is limited
to the river-bed, in that of wind the particles may be
carried along wide belts of country.

Transportation by land-ice. In high alpine districts
and in regions towards the poles, where rain largely gives
place to snow, this snow is in places turned into ice which
moves slowly down from the heights towards the sea.
When the ice is confined to the lower parts of valleys,
these moving masses of ice are known as *glaciers*. We
shall have more to say about moving land-ice in future
chapters; at present it is our object to get a general
insight into the work of the agents of transportation. The
waste which is brought on to the surface of the ice is
under conditions differing from those which prevail where
the material is carried into rivers, inasmuch as the ice is
solid, whereas the river-water is liquid. Accordingly much
of the rock-waste which is transported by glaciers is
carried upon the surface of the ice, but as the ice is
traversed by cracks in many places, some of the material
is lodged in the head of the icy masses or even carried to
its base. A glacier has often been compared to a river of
ice. Like a river it flows down valleys, though with
extreme slowness; accordingly the transportation, as in
the case of rivers, is confined to narrow strips forming the
bed of the glacier. When the ice covers a large tract of

country, as in Greenland, however, the transportation as in
the case of wind may be over large belts.

Breakage of rocks during the processes of transportation.
We have noted that a river owing to its motion is endowed
with a certain power of doing work, and the same is the
case with the moving wind and the moving ice. We have
already regarded some of this work, namely the transportation of material. Now the amount of work which
these agents can perform is not without limit, and if the
river, wind, or ice is supplied with so much material that
its power for work is completely applied to the carriage of
material, it can do nought else. If however some of the
power of the agent is yet available, more work can be
done, and we must now consider the character of this
work.

The work of rivers, wind, and moving ice, apart from
transport, is the further breakage of rock material. This
breakage may occur both in the case of the material which
is being carried, whereby the fragments are rendered
smaller, and also in that of the solid rocks over which
the river, wind, or ice is moving. The latter work is
specially important, as by performing it each of the
agents may lower that part of the earth's surface over
which they are operating. Thus the river and the glacier
may deepen their beds, while the wind may lower the
general surface of the tract over which it blows.

We must eventually consider in greater detail the
type of work performed by rivers, wind, and moving ice,
which is different in the case of the different agents, both
as regards the material transported and the rocky floor
over which the agent is working. At present it is merely
necessary for us to get a general idea of the character of
the work which is done by these agents as a whole.

CHAPTER VII.

THE WEARING ACTION OF RIVERS.

THE ability of a river to wear away its bed, if the amount of rock-waste with which it is supplied is not too great, has already been mentioned. It will be of use now to gain some idea of the nature of the wearing processes brought about by rivers.

If a river is capable of deepening its valley by wearing away its bed, there is no reason why it should not be able to form the valley.

No doubt, when a new land-mass is formed, its surface is not quite even, and the rivers which first course over this land will select the existing depressions in which to flow. These depressions, however, need not be great, and there is reason to believe that they will not be great, for the deposits which are spread over the ocean floor are apt to fill the depressions, and when such deposits are raised to form new land, the surface of that land will be fairly even, unless great rents are made in the earth's crust during the lifting of the deposits above the level of the sea.

That rents may sometimes be formed in this way is probable, and if they did occur, the water would no doubt flow along the lines of the rents. But if we examine the rocky beds of a large number of rivers, we shall find proofs that no such rents have been there. The rocks are often seen to cross the river bed unbroken. Furthermore, the rocks on either side of the valley are often quite level, shewing that the valley is not due to a gaping of the earth's crust comparable to the opening of a book, for if that were so, we should find the rocks on the sides of the valley dipping away from the river. We can to some extent imitate the structure of such valleys if we take a number of sheets of paper, and cut a wedge-shaped portion out of the sheets, as shewn in Fig. 5. This would represent a valley, and the lines around the sides of the valley

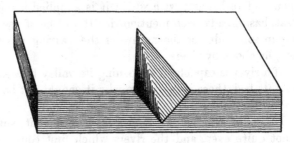

Fig. 5. Sheets of Paper cut to represent a Valley.

would indicate the manner in which the bedding planes of the flat rocks would trend along the valley sides. This structure cannot be produced by bending the paper, or by depressing a portion along fissures, but only by a process of cutting. Similarly when this structure is produced in nature among flat rocks, it can only be due to the cutting away of rocks which once occupied the position which is now held by the valley. The material once extended fairly uniformly over the area, and the missing parts have been cut out by some agent. It only remains for us to consider what this agent was. At one time the material

which once occupied the sites of valleys was believed to
have been removed by the action of the sea. We can
shew that this belief was not correct by two pieces of
evidence. Firstly, the action of the sea, as we shall find
when we consider it, is not of a nature fitting it to produce
ordinary river valleys. Secondly, we have cause to know
that many valleys have been formed during periods when
the sea was far away from them.

The only other cutting agents which break rocks
asunder on a large scale during their movement are
rivers, wind, and glaciers.

Now without entering into the question as to whether
wind and glaciers can produce valleys, it is enough to
note that many valleys have been carved out in districts
where the action of the wind or that of glaciers is not
admissible as a cause for the production of the valleys.

This leaves us to consider rivers as the only other
agent which would be able to carve out important valleys.

A great deal of evidence has been gathered together
by geologists which goes to prove that in the case of a
large number of river-valleys in temperate regions, the
rivers themselves have been the cause of the formation of
the valleys. This evidence is too varied for us to treat of
it in full. It must suffice if we touch upon some of the
more striking pieces of evidence which bear upon the
subject of valley-sculpture.

We will begin by stating that rivers have actually
been observed to form valleys or parts of valleys. Many
cases have been recorded, though the time during which
observations have been made being very short, it is only
very small valleys which have been actually seen in process of formation, and even these are usually formed in
rock of no high degree of hardness.

But though entire valleys have only been formed on a small scale within the memory of man, portions of very important valleys are known to be formed during that period, and as the structure of those portions is quite similar to that of other portions which have been formed in earlier times, it is only natural to infer that the latter, like the former, were due to the same agent; that agent is the river.

It is in the case of waterfalls that this evidence is best grasped, and we may for a moment regard the mode of formation of a typical waterfall such as Niagara.

The actual Falls of Niagara are found to be placed at the head of a gorge which extends for seven miles below the Falls. This gorge is from 200 to 400 yards in width, and about 300 feet in depth in places. The height of the Falls is about 165 feet. The water pours over a mass of hard limestone which has beneath it beds of soft shales, which are easily worn away. The spray from the Falls dashes against these soft shales at the foot, and causes them to crumble. The crumbling is aided by masses of ice being grated against them in winter. By degrees, the removal of the shales causes the limestone above to over-hang, and when a large mass of limestone is left without support, it falls down. The fallen blocks are gradually ground down in the pool beneath the Falls, and carried away in smaller pieces. Thus the Falls must be working up stream.

More than 200 years have passed by since one, Father Hennepin, gave an account of the Falls, with a drawing as they then appeared, and we know that the Falls have worked back since his time, though the actual rate of their retreat is not yet known with certainty. Now the part of the gorge immediately below the Falls is of the

Fig. 6.   A Pot-hole.

same nature as that which is further down the river, and
if the one part has been caused by the retreat of the Falls,
we can hardly avoid the belief that the rest has been
formed in the same way. There is, indeed, much into
which we cannot here enter which confirms this belief.
Now the gorge is over seven miles long, and the river at
the lower end flows into a tract of low ground which is
separated by a cliff from the high ground through which
the gorge has been cut. The Niagara river once fell
over this cliff, and the Falls have cut back for about seven
miles, and to this retreat the gorge is due. We also
know that the cutting of this gorge is, from the point
of view of a geologist, a very recent event, and we are
taught from a study of the Niagara river that when conditions are suitable a river-gorge of some extent and size
may be carved by a river with fair speed.

Many other waterfalls shew signs of retreating in the
same way that the Falls of Niagara are cutting back. In
our own country the best examples are found among the
gently sloping beds of limestone and shale in the Yorkshire dales. Each of these falls occur at the head of a
gorge, which differs from that of the Niagara river chiefly
in point of size.

But though the carving of a river valley is most apparent in the case of a waterfall, we find no hard and fast
line between a waterfall and a cascade, and between a
cascade and a swiftly flowing stream.

In the case of cascades or of rapids, proof of the power
of the river to cut its bed in a downward direction is
often supplied in places where eddies exist. The stones,
gravel, and sand brought down by the stream are whirled
round in the eddying water, and by degrees bore a bowl-
shaped hollow, which is known as a pot-hole (Fig. 6). A

number of these holes may often be seen arranged in a
row, and as they widen, the rock separating two may be
cut through, and the two holes become one. In the same
way others are joined, and one can trace every step in the
process from the making of a row of these holes to the
union of a number of them to form a gorge. This gorge
is clearly due to the cutting action of the river charged
with pebbles, sand, and mud.

But if a river can make a gorge where eddies occur, it
can also do it by sweeping the materials straight along
its bed where eddies do not produce so much effect. If the
current of the stream be strong enough, and the stream
be supplied with sediment to rasp the bed, that bed will
be lowered by degrees, and thus a river-valley grows in
depth. So long as the river has power to rasp, the valley
will be deepened, and if only time enough passes, and
the river has this power, a deep valley may be formed by
the river with no help except the slight differences in
level which determined the course of the river at the
outset.

We have ground then for believing that rivers *can*
cut out their own valleys. But because a thing can be
done in some one way, that is no proof that it *was* done
in that way. Those who were led to believe, from such
observations and facts as we have briefly touched upon,
that rivers could carve their own valleys, were not con-
tent with this belief. They gathered together by degrees
a great deal of additional information about rivers and
their valleys, and found that the new knowledge bore out
the ideas which they had already formed.

One of the most striking bits of proof that rivers do
carve their own valleys is the wonderful likeness of the
minute valleys cut out by the streams on a sloping sand-

bank or mud-flat along the sea-shore at low tide, and the larger valleys which contain the rivers of a big tract of country. Not only do the little tributaries run into the larger streams of the mud-flat in the same way as do the tributaries of large rivers, but the appearance of each little valley-side, with its alternating slopes and terraces, recalls that of a larger valley. As the great river-valleys of that strange region in North America—the basin of the Colorado river—appears to us, so must these little valleys in the mud-flat seem to the little creatures that crawl along their banks. The difference is one of degree, not of kind, and we can indeed trace a gradual passage in point of size between the tiny valley of the sand-bank, and the great gorges of the Colorado, with 6000 feet of slope and cliff from the valley-top to the river-level.

The effect of the work of rivers, so far as we have regarded it, is to produce gorges with steep sides, such as actually exist in many places, notably in the case of the Colorado and its tributaries to which we have just alluded. We must now take into account the ways in which valleys are widened. In so doing we must distinguish between the widening of the upper part of a valley, owing to the slopes of the valley-sides becoming more gentle, and the widening of the actual valley-floor.

With regard to the former change, it is largely due to the work of agents which we have already noticed under the head of weathering. A river cannot for ever saw its way downward; its power of sawing depends on the supply of sediment, the amount of water, and the rate of flow of the stream. The rate in turn depends upon the slopes of the river: still water cannot cut away its bed, there must be some slope to allow of the cutting process. Now, in time a river would saw its way down until the slope was

lessened to so great an extent that no further cutting
down could occur. If the sides of a valley in which the
river had reached this stage were solid rock, they would
still retain their steep slope, but owing to the action of
the weather, the rocks of the sides are loosened, and fall
down the slopes, and this process would go on till the sides
had reached the slopes at which loose pieces would remain
at rest. Further, the little runnels of water flowing down
the valley-sides during and after rain tend to carry down
the loose fragments, and these assist in the process of
making the slopes of the valley-sides more gentle. In
countries, therefore, which have a high rainfall, where
the rivers have sawn their valleys down to a line below
which they cannot go, the valley-sides are marked by
having gentle slopes, and thus the ridges between two
valleys become slowly lowered, and this will go on until
these ridges are lowered to an extent which renders the
slopes too slight to allow of any further reduction of the
height of the ridges.

Another method of widening valleys occurs when
rivers run along the strike of gently inclined rocks of
different degrees of hardness.

Suppose that, as in Fig. 7, there are two masses of
hard rock $H\,H$, with a mass of soft rock S between them.

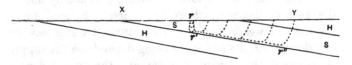

Fig. 7.
Section shewing River cutting sideways between hard and soft Beds.

Let a river be situated on the soft rock, at the point r.
It will cut down in the ordinary way until it has reached

the hard rock at r', a section across its course at different
times being shewn by the dotted lines. When it gets to
r' it may find it easier to cut sideways between the soft
rock and the underlying hard rock, its position at different
times in this stage being again shewn by the dotted lines;
in time it may reach a point r'', the slope on the right of
the figure being due to the undermining of the hard rock
above and the falling of fragments towards the river. In
the meantime the portion of soft rock to the left of rr'
will most likely be washed down, and the valley may by
degrees assume the cross section shewn by a line drawn
through $X\,r'\,r''\,Y$. It will be seen that the river has
shifted sideways, and that the slope on the left of the
stream marked by the upper surface of the lower hard
rock is gentle, while the slope on the right side tends to
be sharper. Not only then do we find valleys widened by
this process, but these valleys are often marked by having
a gentle slope along the dip, and a steep slope facing
the direction opposite to that in which the beds are
dipping. Much of the country in the east and south of
England is marked by valleys of this type, and as the
beds on the whole dip towards the south-east, the steeper
scarps tend to face north-westward.

The widening of the valley-floors is due to the
wandering of streams sideways in certain circumstances.

When a river has ceased cutting downward, if its
course be straight, the stream will simply flow onward,
carrying its materials towards the sea. If, however, part
of a river course be crooked, as shewn by the unbroken
lines in Fig. 8, the river will wander sideways. In a
straight part of a river's course the stream is swiftest in
the centre, for the water at the sides is slackened by

friction against the banks. But when the river comes to
a crooked part, the swifter central current is urged onward

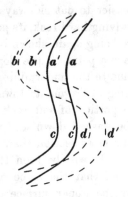

Fig. 8. Plan of River-loops.

for some distance in the same course, and strikes the bank
at a', causing the bank at that point to be cut away. A
backward eddy is set up at the other side of the stream,
a, and the slack water caused by this eddy meeting the
downward flowing water allows pebbles, sand, and mud to
be deposited there, and so the river course gradually
works sideways until it reaches such a position as that
shewn at $b\,b'$. The same process goes on in the opposite
direction at $c\,c'$, causing the river course to travel to $d\,d'$,
and at this stage the course will be as shewn by the
dotted lines.

Owing to this wandering of rivers which have ceased
to cut downward, tracts of river-valleys are widened, and
one river may actually cut its way to an adjoining one
by this process.

As the rivers running along the strike of beds consisting of soft masses lying between harder ones have

often ceased to cut down, we frequently find this wandering going on sideways in the case of strike-streams, as for example in the case of a considerable portion of the Yorkshire Ouse and of the Trent.

From what has been said above, it will be seen that though rivers at first tend to cut up a fairly even surface into ridge and vale, yet, as the processes of wearing go on, and the ridges become lowered, the final result of the wearing action of rivers would be to reduce the surface of the land to a nearly level plain.

There is one special case of river action which must be noticed. It has already been seen that water is capable of dissolving certain substances, one of which is limestone. In districts occupied by limestone the upright fissures are enlarged by this solvent action, and streams often plunge down these fissures, wearing away shafts recalling in some ways those of coal-mines. The water plunging down these shafts often works its way along a bedding plane, and partly by solution and partly by the grinding action of the materials such as pebbles and sand which it carries down, wears away a course for itself underground. In this way are produced the caverns which occur with such frequency in limestone districts, as in the dales of West Yorkshire and Derbyshire, and in the Mendip Hills.

We must now say a few words with regard to the changes which take place in the fragments which are carried down by the rivers. These as well as the rocks of the river-bed are worn by the stream. The fragments produced by weathering are chiefly marked by having sharp edges and corners. As these fragments are knocked one against the other and are pounded against the rocks of the river-bed, the edges and corners become blunted, and the larger the fragments, the more worn they become. Thus the large pieces carried by a swift stream when rolled along the river-bed for some distance are made into rounded pebbles. They are not often round like a ball, but usually every sharp edge and corner which they once had is worn away. Let the reader look at a handful of pebbles from the bed of some brook, and note how they differ from the fragments which lie at the foot of a mountain cliff. Besides being rounded, the former are on the whole smaller than the latter. The smaller grains, forming sand, are not worn to the same extent as are the pebbles, but even here the sharper angles and edges are by degrees rounded, and it is only the very fine particles of mud which tend to retain their original sharpness.

Special attention is called to the condition of the pebbles and sand grains which have been borne onward by rivers, as by their appearance we are able often to distinguish them from those which have been carried onward by wind and by glaciers.

CHAPTER VIII.

THE WEARING ACTION OF WIND.

WE have paid special attention to the work of rivers, for its effects are at the present day very wide-spread, being seen in tropical and temperate regions: we have, also, no reason to doubt that in former times large portions of the land-tracts were worn away by river action.

The work of wind and glaciers is most marked, however, as it was no doubt in past times, in special regions, the work of the former being mainly important in desert tracts, and of the latter in arctic and Alpine areas.

But we must nevertheless pay some attention to the work of wind and of glaciers, for, as we shall eventually see, our own country has in the past been subjected to climatic conditions differing from those which at present affect it, so that at one time the site of Britain was occupied by arid desert, at another by masses of land-ice creeping from its mountain-tracts.

In desert regions, as already stated, the rocks are largely weathered owing to lengthening by heat and shortening by cold. The heat by day is often very great, and owing to the clear skies, the nights are often very cold. Accordingly the solid rocks which are bared in these tracts often break into pieces of various sizes, and the rate at which the weathering takes place is in many places rapid.

The fragments which are broken from the solid rocks as the result of weathering become smaller and smaller by undergoing changes like those which caused their separation from the parent rock, until by degrees grains are produced which are so small that they can be carried along the floor of the desert, or whirled in the air by the winds which blow, often fiercely, over these desert tracts.

The wind is not confined to a narrow strip like that which forms the bed of a river, but acts along wide belts of country. It rubs the grains of sand and bits of dust against the rocks which form the floor of the country over which it blows, and as the result of this action these rocks are all by degrees worn away.

We often see glass bottles with letters cut in the glass. These letters may be cut in more than one way. They may be scratched by a diamond : or again a certain acid affects glass, and by covering the glass with some substance which is not affected by the acid, and then removing it from the parts which we desire to cut, and applying the acid, the latter eats into the glass. But there is another way, and it is that which concerns us, as throwing light on the action of sand in deserts. If those parts of the glass which are not to be cut are covered, and if instead of treating the uncovered part with acid, a current of air is made to blow sand against the glass, this sand can be made to cut the letters. We speak of this air-current charged with sand as a *sand-blast*; now the sand-laden winds of desert regions form a natural sand-blast, and the rocks are by degrees fretted away by its action. If the rocks of the desert floor were all equally hard, and the sand-blast worked with equal power over the whole of that floor, the rocks would be rubbed away at the same rate over the whole tract.

The rocks, however, are affected by the sand-blast in different degrees, and accordingly parts of them are worn away more rapidly than others; the harder rocks tend to be worn away more slowly than the softer and stand out to a greater extent after the sand-blast has been in action for some time. This is well shewn in the case of the celebrated Sphinx which stands by the Pyramids of Egypt. The Sphinx was carved from a rock which consists of bands of different hardness, and having been subjected to the sand-blast since it was formed the hard bands project as ridges, while the softer have been furrowed.

When the rocks are of a particular nature the effect of the sand-blast is to give rise to a polished surface, sometimes making the rocks appear as though glazed, and this surface is different from that which is produced by running water charged with sand.

The various planes of weakness which rocks present, such as bedding-planes and joint-planes, assist in causing the rocks to be worn away at unequal rates in different parts. Those portions of the rock which lie along these planes are often worn away more rapidly than the parts which lie away from them. Thus the joints may after a time be marked by wide fissures, and the position of bedding planes by deep grooves. Along cliffs, the wearing along joints may end in the separation of large blocks from the main cliff. These blocks will stand out as pillars. As the sand works at the base more than at the summit, the pillars are by degrees cut away from below, and they are sometimes seen forming mushroom-shaped masses with slender stalks. At last the stalks are cut through, and the summit of the pillar topples over, and is still further worn away.

There are other effects at work in deserts into an

account of which we cannot enter. These to a further
degree tend to cause the rocks of uneven surfaces to be
unequally affected by the sand-blast. Accordingly, one
of the noteworthy features of a rocky desert is the
fantastic way in which the rocks are carved into columns,
pinnacles, and mushroom-like masses.

We must now refer to the way in which the fragments
are changed while borne along by the wind. The larger
pieces which are broken off by the weather are usually
angular, and the smaller bits are at first angular, but
when carried by the wind these by degrees become
changed in form.

The sand-grains hurried along by running water are
to a certain extent buoyed up by the water, and only
now and then do they collide with each other, or are
knocked against the rocks of the river-bed.

When grains are carried by the wind, they are not
buoyed up by the air to the same extent as by water, and
are more frequently knocked against the rocky floor, and
collide with one another. As a result of this they are
worn smaller than the water-borne grains, and the angles
and edges are knocked off to a greater extent. It is
found therefore that, on the whole, the sand-grains of a
desert tract are smaller and rounder than those which
have been carried by water and laid down in a water-
tract. In many cases the grains are so well-rounded
that each grain approaches in shape to a little globe.
These grains have been compared to millet-seeds, and the
sands which are formed of such grains are spoken of as
millet-seed sands.

The discovery of millet-seed sands among rocks formed
in the remote past is one of the bits of evidence upon
which we rely when we infer that certain regions have
been deserts in past times.

Fig. 9. A Contraction, Morainic

Fig. 10. Rocks rounded and smoothed by Glacier.

Fig. 9.   A Glacier with Moraines.

Fig. 10.   Rocks rounded and smoothed by Glacier.

CHAPTER IX.

THE WEARING ACTION OF GLACIERS.

IF we take our stand upon some hill-top in Britain we may notice the river at the bottom of a valley at our feet winding about like a silver ribbon. Let us now suppose ourselves standing on some high eminence in Switzerland. Beneath us there may be no river, but a white mass of ice, marked here and there with long strips of darker material, may be seen occupying the valley-bottom. It looks somewhat like a river, and has indeed been spoken of as a 'river of ice.' Like the British river, it may be seen to have a winding course, though the bends are usually fewer than those of the stream of water. Such a mass of ice or *glacier* as it is termed is shewn in the picture (Fig. 9).

A glacier has many other resemblances to a river besides the general one noticed above, but it differs in that it is composed of ice and not of water.

What, then, is this glacier, and why is it found in many upland districts?

The hills towards the source of the glacier are found to be largely covered with snow in summer as well as in winter. The snow comes down to about the same level along all the hill sides of any particular district; the level is that at which the air is warm enough to cause

4—2

the snow to melt. It will be noticed, however, that glaciers often descend far below this level, and this is a matter which is important to us.

If we can manage to reach the snows among the high hills, we find that parts of the snow are unlike the powdery snow which falls in England in the winter months. The surface may resemble that snow, but if we look into a fissure in the snow (and there may be many), the snow below the surface is different from that with which we in England are familiar: parts of it may recall the snow of a snowball, and indeed it is to some extent similar because it is due to the same causes. The snowball possesses its characters because it has been squeezed, and while being made parts of it melt and become frozen again, binding the mass together, until at last the snow-ball becomes like a mass of *ice*. The snow of the higher mountains is also gradually changed into ice, and we can trace the gradual change from the fine, freshly fallen snow of the surfaces of the *snow-fields* (as the tracts of snow in upland regions are termed) into the ice of the glaciers.

Here the comparison with the snowball ceases. If the snow were simply changed into ice, like that of the snowball, the glacier would end at the same level as the snow itself does, whereas, as we have seen, many glaciers end much below the *snow-line*, as the lower limit of the snow is termed. This is because the ice of the glacier *moves*, strange as it may appear. *How* it moves is a question which has tasked the minds of many men of science, and it is a very difficult question, into which we must not enter: it is sufficient to know that it *does* move. The fact that the ice of glaciers moves onwards can be proved, and has often been proved, by very simple observations, to which we must now refer.

If a row of stakes be driven into the glacier from one side to the other, so that the line passing through the stakes is at first straight, and if two stakes be placed on the rock, one at each side of the glacier, to form the ends of that straight line, and if the stakes be again observed after some time has passed, it will be found that they have been moved, so that the stakes on the glacier are now further from the source of the glacier than they were when driven in. This is proof that the glacier moves. But a further important fact will be noticed, namely, that the line joining the stakes on the ice is no longer straight, but curves in a form reminding one of a bow, and the convex side of the bow faces the lower end of the glacier while the concave side looks towards its source. From this we learn that the central part of the glacier moves faster than the sides, as does the central part of a river, and for the same reason: the rocks of the glacier's bed act as a drag, and so the sides in contact with the rock move more slowly than the central part, which can move more freely.

This movement of ice seems strange, but it is not unlike the movement which may be seen taking place in a mass of pitch placed upon a slope; and whatever be its cause, there is no doubt as to its occurrence.

But though the ice of a glacier, like the water of a river, moves onward, the rate of movement in the two cases is very different. The glaciers of Switzerland move at the rate of a few inches a day, a movement of two feet in that time being unusual. It is true that parts of the moving ice of Greenland have been found to travel much faster, the ice in some places having been observed to move at the rate of over fifty feet in the twenty-four hours, but this is in special circumstances.

Of the various features presented by a glacier, those due to the movement of the ice are of great importance to the geologist. This movement causes the great cracks which extend downward from the surface of a glacier, which are known as *crevasses*. They form a very real danger to those who walk over the ice of glaciers, for when the ice is covered with snow they are often unseen, and the unwary traveller may fall into one. And as they may entrap a man, so they may and do swallow up stones and other materials which previously lay upon the surface of the ice.

Some of these cracks are due to the greater speed of the centre of the ice than that of the sides, while others are due to differences in the width of the glacier's bed, and especially to changes in the slope of that bed. When a river plunges over a cliff it forms a waterfall; when a glacier does the same it gives rise to an *icefall*; at these icefalls and a little below them the ice of the glacier is usually broken by numerous fissures between which a bristling array of ice-pinnacles renders progress on the ice very difficult.

Below the snow-line the less fissured parts of the glacier are often furrowed by water-channels, the water being supplied by the melting of the surface of the ice, and the channels in summer may be occupied by rivulets, which flow on till they come to crevasses. On reaching the fissure the stream plunges down and works a circular shaft in the ice till it reaches the floor of the glacier, when the water flows on between the ice and the glacier's bed, so that, at the end of a glacier, the water due to the union of the various streams which have plunged down these shafts and have reached the lower part of the glacier in other ways, issues from the ice, often as a river of some size.

We are now in a position to consider the work of the glacier which is of importance to the geologist.

When taking account of the effects of weathering, we had reason to note that in cold districts the effect of frost is to break off angular pieces of rock, and if the slope is great enough these fragments work their way towards the valley-bottom. If that bottom be occupied by a glacier the fragments will collect on the glacier's side, to form a fringe of loose blocks,—loose, save that they may be frozen together by the ice of the glacier itself. As the ice moves on, each part of the glacier's side is brought opposite to the points where the supply of material is abundant, and when laden with the stones, it moves past this point, which is now supplying fragments to a new part of the glacier. So by degrees a fairly regular fringe runs all along the side of the ice, forming what is known as a *moraine* : each side of the glacier will then come to possess a side-moraine. When two glaciers join to form one, as they frequently do, the side-moraines which come together in the central part of the main glacier will form the central moraine, and if a glacier be formed by the union of a number of tributaries, we may and often do find several of these moraines running as ridges down the central parts of the glacier (see Fig. 9).

As the material which is borne on the surface of the glacier is carried with extreme slowness, little or no change takes place in the appearance of the fragments after they have been carried for long distances :—they may move onward for years still keeping unchanged the edges and corners which they possessed when they fell upon the ice.

These moraines which are formed on the upper surfaces of glaciers may be spoken of as *surface-moraines*.

Let us now turn to regard the changes which take place in any material which is carried below the ice between the glacier and its bed.

Here we have to deal not only with onward carriage of the material, but with change in the character of the fragments during that carriage, and also with change in the appearance of the rock of the glacier's bed, for carriage below the ice is accompanied by wearing away of rock.

We have seen that fragments fall down the fissures, and are carried down the shafts worn by water falling into these fissures. These fragments become wedged between the ice and the bed of the glacier, and as they are carried on by the glaciers they knock off other pieces from the rocks of the bed which are in turn borne onward.

The materials thus wedged into the ice are moved in a way differing from that which marks the onward sweep of the substances carried by a river. The latter are allowed a certain freedom of motion. They are rolled round and round, bumping against the river-bed. The fragments beneath the ice, however, are more firmly embedded in the ice, and as they move slowly forward are pressed down by the weight of perhaps hundreds of feet of ice lying above them.

The lower surface of the glacier, charged with fragments of material of various sizes, from large pieces of rock many inches in diameter to particles of the finest mud, may be compared to a gigantic rasp moving over the rocks of the glacier's bed, and parts of these rocks are therefore rasped away, thus causing the bed to be to a greater or less extent deepened. There is much difference of opinion as to the power of a glacier to deepen its bed,

though granted a sufficient lapse of time, the deepening
in the case of a glacier, as in that of a river, must go on
until the conditions are such that the glacier can cut
down no further.

More interesting to us however than the actual amount
of rasping work which the glacier performs, is its nature;
we wish to know how the erosive action of ice differs
from that of rivers on the one hand and that of wind on
the other. That fragments of rock are torn away—the
ice taking advantage of the planes of weakness such as
joints—is certain, and a good deal of the wear produced
by glaciers is no doubt caused in this way, but this is not
the type of ice-wear by which ice-action may be most
readily recognised.

Owing to changes in climate glaciers sometimes advance
and sometimes retreat. If the rocks at the end of a
glacier be examined during a period of retreat, those
which have recently been covered by ice may be noticed
to possess certain features. In the first place, the corners
and edges of projecting rock masses are not only knocked
off, but the general surface of these projecting masses is
smoothed and rounded off, as if by a curved plane,
having its concave side facing downward, and this
smoothing process is specially noticeable in those parts
of the projecting rock which face up valley, that is toward
the direction from which the ice comes, for the parts
facing down valley usually remain rough, and often shew
signs of tearing. Owing to this rounding and smoothing
the rocks thus affected have been fancifully compared to
flocks of sheep lying down. An illustration of some of
these rounded rocks is given in Fig. 10.

But in addition to this, if the rock be of a favourable
nature the fine mud carried by the ice acts as a polishing

powder, and the smoothed surfaces often display a considerable degree of polish, differing from the mere smooth-
ness of water-worn rocks, though in some degree comparable
with the polish of rocks which have been affected by wind.

Furthermore—and this is of particular interest—the
hard and sharp bits of rock, both pebbles and grains of
sand, are pressed against the surfaces of the rocks over
which the ice passes, and grooves or scratches are
produced on the surfaces of the solid rocks in this
manner, just as one can scratch a table by pressing a
grain of sand against it. These scratches are formed in
parallel lines, the lines running in the direction in which
the ice moves.

Rounded, scratched, and polished hummocks of rock
are then very typical of the action of glaciers, and the
former occupation of a district by glaciers may be pretty
confidently asserted when rocks affected in this way are
found in the district.

But, as in the case of work by rivers and wind, so in
that of ice, not only are the rocks of the bed changed,
but also the fragments which are carried over that bed.
By degrees the fragment which fell down a fissure with
all its corners and edges in the condition in which
they were when the piece was broken from its parent
rock becomes modified as the result of pressure against
the glacier's bed, and the same is of course true of the
pieces which are knocked from off the glacier's bed
during the onward progress of the ice. The larger
fragments have their edges and corners blunted without
being actually rounded off like those of the pebbles
which are worn by water action. The mud which is
also carried by the ice acts as a polishing powder on
many of these fragments just as it does on the solid rocks,

so that when the stones are of suitable nature, they have a polish imparted to them resembling that which exists on many of the hummocky rocks over which the ice has passed. As these fragments are carried onward, gripped firmly in the ice, they are often pushed over projecting grains in the rocks beneath, or over grains lying between rock and ice, or the grains may be in the ice and travelling faster than the larger fragments. These grains produce scratches on the larger pieces of stone, resembling those which occur on the rocks over which the ice has passed, except that they more frequently run in directions which are oblique to each other. These scratches are usually parallel to the direction of greatest length of the stone, for the stone travels most easily with its greatest length parallel to the direction of movement of the ice. Now and then, when the stone meets some obstacle, it is turned, and travels onward for a short time with its greatest length oblique to the direction of movement, and accordingly some scratches are made obliquely across the stone, but these are usually few in number as compared with those which run in the direction of greatest length.

A fragment such as we have described, with its edges and corners blunted, with a polish on its surface, and with this polished surface marked by scratches, is one of the most useful indications of glacial action. It is true that such a fragment may be to some extent imitated by pieces of stone whose features have been caused otherwise, but in general it is easy after a little experience to distinguish between these glaciated stones and others which in some degree resemble them.

Fragments which have been carried by ice are known as *glacial boulders*, and of these boulders we have seen that those carried on the surface of the ice differ from those borne below, inasmuch as the former are unchanged, while the latter have been modified in the ways described above.

The reader must understand that all three kinds of effects are only produced when the rock is of a suitable nature. Usually in a mass of glacial material which has been borne beneath the ice a large number of the boulders are blunted only, some may be in addition scratched without being polished, and a few may be polished also.

The materials which are worn from the edges and corners of the larger fragments are carried onward as grains of gravel and sand and particles of mud. Owing to the great weight of the overlying ice these smaller fragments are frequently crushed into exceedingly fine particles of mud : the fineness of the mud may be judged by examining that which gives the milky appearance to the stream which issues from the lower end of a glacier.

It has been stated that owing to change of climate the end of a glacier is not always in the same place, but it may nevertheless remain at one place for a number of years. At the end of the glacier the materials which have been carried below the ice (as well as any which have been borne in the mass of the glacier) are laid down, and mixed with the substances which were carried on the surface of the glacier. All of this loose material, boulders, gravel, sand and mud, builds up a *terminal moraine*, which often forms a crescent-shaped ridge with the hollow of the crescent pointing up the valley.

We have considered some of the effects of glacier action by taking into account the work of such glaciers as may be found in the upland regions of Switzerland and Norway. It must not be supposed, however, that these

comparatively small masses shew all the work which
can be performed by land-ice. In Alaska a number
of upland glaciers join on the lower ground to form a
great tract of ice, part of which is slowly rotting
where it lies, so that one may actually find forests
growing on the moraine-stuff, with perhaps a thousand
feet of glacier ice beneath it. In the interior of Greenland
the whole country, hill and valley alike, is covered by ice,
which here forms a great ice-sheet, moving from the
interior seawards, and many features are displayed by the
ice-sheet which cannot be studied in the case of ordinary
glaciers of the Alpine type. Lastly, there is the great
mass of ice of Antarctic regions, of which the outer edge
only has been touched. Here the work of land-ice is
probably being carried on to an extent without parallel
in any other part of the world.

When parts of these moving masses of land-ice reach
the sea fragments of various sizes break off and float
away as icebergs, bearing with them some of the materials
which were included in the ice before it became detached.
As these icebergs float into warmer regions they gradually
melt, and the result is that materials of the nature which
we have described as being due to the action of land-ice
may be deposited upon the sea-floor in addition to being
left stranded on the surface of the land. Much of the
material brought south by the icebergs which have
broken away from the Greenland ice-sheet is deposited
upon the banks of Newfoundland, which have been somewhat fancifully termed "the terminal moraine of the
Greenland ice-sheet," though of course only a portion of
the material is there stranded.

CHAPTER X.

THE WEARING ACTION OF THE SEA.

WE have noted in the foregoing chapters how various agents which are at work over the surface of the land-areas tend by degrees to reduce those areas down to a definite level by erosion. While these changes are taking place over inland tracts destruction of the land is also going on along parts of the coasts owing to the action of the sea.

The agents whose work we have been considering cut downward, whereas the sea chiefly acts sideways, its downward action being very limited.

The sea is largely potent as an agent of destruction by making use of the broken fragments of rock, which when urged against the rocks of the land wear them away. We have compared the action of a glacier to a rasp, that of the sea is more aptly likened to a battering-ram.

The waves which are caused by wind blowing over the surface of the sea are mainly responsible for the destructive action upon the land, while the currents which affect the upper waters of the ocean act as carriers, causing the material derived from the destruction of the land to be carried in fragments from place to place.

It has been found as the result of observation that

the action of the waves is confined to the upper parts only of the oceans. At a depth of a few dozen feet from the surface the movement of the waves is unfelt, and accordingly the battering of rocks by wave action is restricted to the sea-coast and to the narrow strip of sea lying off the coast which is not too deep to be affected by wave action.

Again, though currents produce marked movement to a greater depth than do waves, there is little movement due to these currents at a depth greater than about 150 feet.

The action of the waves in causing wear of the coasts is most readily studied where cliffs fringe the coast line and a beach lies at the foot of the cliffs.

Part of the loose material of the beach may be derived from the cliff above by the action of the weather upon the rocks of the cliff, and addition is made to this owing to the actual ravages of the waves. During storms the waves hurl the shingle and sand against the rocks of the cliff, and batter away portions of these rocks. Thus by degrees the cliff tends to become undermined, though it is not common to see the upper part actually overhanging, for owing to landslips, the action of the weather, and other processes, the upper part of the cliff also gets worn away, and most cliffs actually slope outward towards the bases.

Thus in course of time the cliffs are worn further and further back, and the sea gradually encroaches upon the land, while the material which is derived from the destruction of the cliffs is carried out to sea, to be there arranged in a way which we shall describe in a later chapter.

The rate of destruction varies with the nature of the

rocks, the roughness of the sea and other things. On some parts of the eastern coast of England the sea has eaten away the land at the rate of many yards in the course of a year, so that tracts which were land in Roman times may in some cases be under the sea at a distance of considerably more than a mile from the present shore. Though, as stated, the action of the sea is most readily seen where cliffs occur, the destruction of the coasts of low-lying tracts may take place without the formation of cliffs.

When we study the action more closely we find that, as in the case of the action of wind, rivers, and glaciers, the planes of weakness in the rocks have their effect upon the nature of the erosion. The waves tend to cut away the rocks along these planes, giving rise to caverns and arches, and converting projecting ridges into isolated stacks and needles, by cutting through them. These features may be seen along many parts of our coast, as at the west end of the Isle of Wight, where the 'Needles' are well-known.

On the whole the soft rocks are worn more readily than the harder, and the latter tend to stand out as capes while the former are worn into bays. This is well seen along a great part of the coast-line of England and Wales. The sea-lochs of Scotland however are due to a different cause, as are many of the estuaries of England and Wales. This will be considered later.

The effect of the sea upon the loose materials which are worn from the coast must now be considered. If the cliffs sink into deep water, the fragments broken from them may fall below the level at which wave-action is of importance, and little further change of the fragments will occur. This may also be the case in sea-lochs with high land around, where the waters may remain almost unruffled, when the sea outside is lashed into fury.

When the materials form a beach between high and low tide-levels, the fragments are ground against one another, especially during storms, and the angular bits of rock become rounded, so that one usually finds well-worn pebbles on the beaches, with edges and corners completely rounded off. The finer particles exist as grains of sand and mud, which do not differ in any important way from those formed by the action of rivers.

The onward carriage of the materials away from the coast will be more conveniently taken into account when we discuss the way in which deposits are formed, but we must here note that there is an on-shore movement of beach materials. If the wind blows on to the shore from off the sea in a slanting direction, a current of water will be set up along the shore, and if this be sufficiently strong it may move pebbles along the shore. During this onward movement the pebbles become further worn and rounded, and they tend to accumulate in sheltered bays, and sometimes to build up shingle-spits across the mouths of bays.

It must be remembered that while the action of the sea is assisting to reduce parts of the land along the coast by its wearing action, the other agents, whose work we have already regarded, are working inland. We have seen that these alone, if allowed to work undisturbed for long ages, would reduce the lands to a nearly level surface, not far above the level of the sea. But the sea would not allow the tract to rest undisturbed: it would further lower it until everything was destroyed above the lower limit at which wave-action is able to work. Hence, as the combined result of all agents of wear, if a sufficient

length of time elapsed, and nothing else took place to
modify this result, all land tracts would be reduced to a
flat surface, some feet below the level of the ocean-surface.

Such planes have been formed again and again in past
times. If the rocks which compose them be aqueous
rocks which have inclined bedding-planes, these planes
will abut against the plane produced by erosion, as shewn
in the lower part of Fig. 11 (see also Fig. 3, p. 16). In

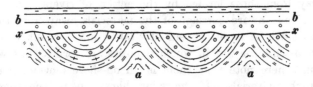

Fig. 11. Section of an Unconformity.

a a Older Rocks.   *x x* Plane of Unconformity produced by Erosion.
     b b Newer Rocks laid down after the Plane was formed.

this figure, which is an imaginary section (or vertical
cutting), the curved lines indicate the bedding planes of
the rocks which have been worn level to the line *x x* by
the agents of erosion.

Now if new deposits be at any later time laid down
on the plane thus formed, their bedding planes will lie
parallel to the plane *x x*, as shewn at *b b* of the figure.
Such an arrangement is termed an *unconformity*. The
occurrence of an unconformity is very important to
geologists. It implies that the area was sea when the
rocks *a a* were formed (if they are ordinary sea-deposits).
After this the tract must have been *raised* to allow the
agents of wear (which as we have seen only work upon
the land or in very shallow water) to plane down the

rocks, and finally, it must have been *lowered* to allow the deposition of the new set of beds, *b b*.

If we consider the nature of the various agents which cause rock-erosion we shall find that they are able to perform their work because of the sun's heat. The changes of temperature which split the rocks into fragments are directly due to the sun's heat. The rain which assists in this work is due to evaporation of water by the sun and its subsequent condensation. This rain gives rise to rivers, and the snow due to condensation of the vapour in cold regions to glaciers. The wind, which operates especially in desert regions, is due to difference of temperature of different masses of air, caused by unequal heating of those masses by the sun. Lastly, the wind causes the ocean waves which erode the coasts, and also the principal currents which carry the material.

The sun, then, is the source of *energy*, by which in conjunction with the action of *gravitation* the work of erosion is carried out.

CHAPTER XI.

ACCUMULATIONS FORMED ON LAND AREAS.

In the preceding chapters the various ways in which the rocks of the earth's crust are broken up and the fragments carried from place to place have been considered. It now remains for us to pay attention to the methods by which these fragments are rearranged to form new rock-masses.

This part of our study is of special importance, for as the stratified rocks of the past may be compared to the leaves of our geological volumes, furnishing as they do when rightly studied and interpreted the records of earth-history, it is very necessary that we should know how rocks similar in most respects to those which were laid down in past times are being formed at present.

The loose materials, when travelling onward, are in the main carried down-hill. Here and there grains of sand may be carried up slopes by the action of the wind, but even in the case of wind the movement is on the whole in a downward direction. The great agents of transport—the rivers—bear their burdens ever downward, and, with local exceptions, glaciers also do the same.

The particles and fragments may halt on their journey for long periods of time, but the interrupted journey downward will, if no other change occurs, be eventually resumed, and there is no proper rest for the products of erosion until they have reached the sea and been lodged upon its floor. There they may repose until other changes, the nature of which will be considered in a later chapter, may cause their uplift, when they may once more form parts of land.

As the land is on the whole an area where destruction of rock is going on, and the sea is the great receptacle of the broken materials, it follows that the bulk of aqueous rocks which were formed in past times were laid down upon the sea-floor, and we shall therefore have to pay careful attention to the characters of these *marine deposits*.

The comparative rarity of land accumulations among the older rocks renders their occurrence a matter of exceptional interest, and we may therefore, before describing the modern marine deposits, devote our attention to a brief account of the accumulations formed on the land masses, whether on the actual land surface, beneath expanses of fresh water, or in inland lakes formed of salt water.

We will commence by an account of those materials which are accumulated on dry land.

In the first place, we will notice those accumulations which are due to the weathering of the rocks, without any transport of the broken particles from the place where the weathering occurred. Of this nature are many *soils*, though the accumulated material need not be of a nature suitable for growth of plants.

In places where cultivation has not produced change in the soil, a gradual passage can be traced from solid rock below to soil at the surface. The rock some way down is seen to be undergoing breakage along planes of weakness. Higher up, the fracture has gone on to such an extent that the larger rock fragments are actually separated from one another, the spaces between being filled with the smaller particles due to the acts of breakage. Still higher, the amount of small material becomes more abundant, and the size of the larger fragments correspondingly less. Here we may find the larger roots of plants mixed with the broken rock, and by their decay contributing some organic matter to the accumulation. Still higher up the inorganic materials become more finely divided, the process of division being here often assisted by burrowing animals such as worms, and the amount of vegetable matter mixed with these materials becomes greater, until near the surface we meet with an ordinary soil composed of varying amounts of inorganic and vegetable matter thoroughly mixed together by the action of worms and other creatures.

These accumulations differ according to the nature of the rock from which they are derived, containing variable proportions of lime, iron, and other substances, as the rocks beneath vary. They also differ according to the proportion of inorganic and vegetable substances which go to form them. In desert regions we may have accumulations devoid of vegetable matter, while in flat swampy tracts of humid regions the amount of vegetable matter may be so great as to give rise to peat, and between these two extremes we can trace every intermediate stage.

In these soils we often find relics of various dates mixed together. Modern and medieval pottery may be found in Britain together with that of the Romans, and even with relics of earlier date, when the inhabitants of this island were not able to make metal implements, but fashioned them from stone. It has been urged therefore that no important wearing of rocks has taken place during long ages where soils of considerable antiquity are found. But although the soil has been there for long periods, and the objects of considerable size have remained embedded in it, the particles of soil are ever changing. The skin of our bodies is of the same pattern through our lives, but it is ever being worn off the surface and replenished beneath. Similarly the particles of soil are washed or blown away from the top, but what is thus removed is replaced by other particles worn from the rocks below, and thus the soil must gradually be lowered, owing to the wear of the underlying rocks.

Of accumulations formed of weathered fragments which have travelled to a greater or less distance from their parent rock, we must notice the more important types.

Firstly, there are fragments which simply fall down slopes and accumulate at the base. Of this nature is the rock-waste formed at the bases of cliffs, which was briefly noticed in chapter v. As little further change takes place after the fragments are split off they retain the angular outline which was determined by their fracture along planes of weakness. These accumulations are frequent in mountainous districts, and are termed *screes*.

Next we may notice the materials which are washed down slopes by occasional runnels of water which course down such slopes during heavy rain. These runnels sift the finer particles of the upper parts of the slope, and when the runnels reach flatter ground at the base of the slope, the fine particles are deposited, often in the form of a mass of muddy material of a fertile character. These deposits are spoken of as *rain-wash*.

The accumulations formed from wind-borne material

are best seen in deserts where, as already explained, wind-action is specially marked. In countries with much vegetation, the roots of the plants bind together the particles of the surface, and the action of the wind is not very pronounced. But in temperate countries like our own there may be tracts where the wind can find material in a state fit for removal, namely on sea-coasts, when the winds blowing from the sea can remove the loose sand grains from the shore and bear them inland. Here we often find belts of sand-hills or *dunes* formed by the piling up of wind-borne sand grains. In the sheltered tracts of desert regions these accumulations of wind-blown material may be piled up to a very great thickness, the upper parts being marked by dunes. The nature of the particles of which these blown sands are formed has already been described when discussing the erosive action of wind. They are often millet-seed sands.

The deposits formed by rivers are of interest in many ways. Rivers are capable of carrying material, and of eroding their beds. They cannot do an unlimited amount of work, and the whole energy of a river (that is its capacity for doing work) may be taken up in the work of carrying its load. This energy depends upon the amount of water and the swiftness of the stream. If the river in one part of its course has its swiftness lessened, it may be unable to carry so much material as it was able to bear higher up when its rate was more rapid, in which case some of the material will be deposited on the river bed, and the bed will by degrees be built up. This building up of a river bed takes place most frequently, though by no means exclusively, when the course of the river is through flat country.

In the flat parts of a river's course during periods of

flood, the river often overflows its banks, so that a wide expanse of water may occupy the valley floor. This overflow water often moves onward but slowly, and accordingly the sediment which it contains will be deposited on one or both sides of the actual river bed. Not only is the river bed gradually heightened then, but the ground on either side of the river is also raised by deposit. This deposit is known as river *alluvium*, and the level tract which it forms is spoken of as an *alluvial flat*, or if very extensive, as an *alluvial plain*. The materials of these plains vary in degree of coarseness, according to the conditions of their formation, but they are frequently composed of fine mud, even though coarse shingle and gravel may be deposited in the actual river bed.

Now changes may occur which cause the energy of a river which has been depositing alluvium to be increased, and it will then cut through its alluvial plain, and down into the underlying rock. If the new part of the valley is narrower than the older part, as it usually is, parts of

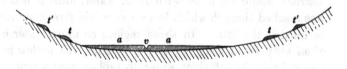

Fig. 12.

v River.   *a a* Modern alluvial flat.   *t t* Lower terraces.
t′ t′ Upper terraces.

the old alluvial plain may be preserved as level strips at some height above the existing river; these are *river terraces*. Alternate periods of erosion and deposit of alluvium may occur more than once, in which case we may find two or more sets of terraces one above another, as shewn in Fig. 12, which is a section across a river valley, with a modern alluvial flat and two sets of river terraces.

River courses are sometimes interrupted by the occurrence of one or more lakes. As the river current is checked when it enters these lakes, deposits are formed at the river mouths as deltas. These deltas gradually grow outward, until the lake is finally silted up, giving rise to a tract of alluvium, through which the river will then wind. As the mode of deposit of these sediments in lakes is generally similar to that which occurs in the open ocean no further description need be given here.

The deeper parts of the lakes may be so far from shore that all the sediment is deposited before it reaches these parts. If living creatures exist on the floor their shells and other hard parts may form deposits of organic origin. Such are the shell-deposits which are now being formed in many lakes. These again are akin to deposits being piled up on parts of the ocean floor.

But there is one kind of lake deposit which deserves special mention.

In desert regions the hollows which in more rainy districts would be filled with fresh water, until a notch was reached through which the water would flow, may not be filled to the brim. In these regions so much water is often evaporated that the rivers entering the hollow no longer bring in sufficient water to replace that which is carried away in the form of vapour. Now the water which is evaporated is fresh water. If water holding any kind of salt in solution is evaporated the salt is left behind, as can be readily seen by dissolving some ordinary table-salt in fresh water, and then boiling the water. Before all the water is boiled away some of the salt will be deposited in the solid state.

The waters of these desert lakes have various substances in solution, such as carbonate of lime, sulphate of lime or gypsum, and chloride of sodium which is the common salt that we eat. Accordingly, as evaporation goes on layers of these and of other substances which we have not noticed will be laid down on the lake floor; first, the least soluble, and lastly, the most soluble. These deposits, known as chemical precipitates, are very common in the lakes of desert regions, though they may be formed elsewhere on a smaller scale, as, for instance, in caverns where carbonate of lime is deposited during evaporation of water in the form of pendent stalactites from the roof, or sheets of stalagmite on the sides and floor of the cavern, and also where springs containing much carbonate of lime in solution gush out of the ground.

The last accumulations to be noticed as being formed on land areas are those due to moving land-ice, but it was found convenient to notice the main characters of these when describing the erosive action of glaciers, and we need add nothing further here to that description.

CHAPTER XII.

DEPOSITS FORMED IN THE SEA.

WE have regarded the methods of carriage of the waste of the land in a seaward direction. The principal agents of transport—the rivers—bear the material onward, partly as visible fragments suspended in the water or pushed along the bottom, partly invisibly in a state of solution. Both kinds of waste may and do contribute to the formation of deposits on the sea floor. We will begin with a description of the modes of accumulation of that part of the waste which has been carried down in a visible condition, remembering that besides the materials brought down by rivers, other fragments have been produced by the action of the sea-waves, as described in chapter x.

It was seen that the carriage of rock-fragments depended partly on the rate of flow of the water. When a river enters a lake or the sea its current is checked. The water does not suddenly cease flowing, but is first slackened, and continues moving more and more slowly, until at last the movement ceases. Now the swifter the stream the larger the fragments which it can carry, so when the current slackens the largest pieces are first dropped, then the smaller ones, which are therefore carried further away from the shore than those of greater size. The water therefore acts as a kind of sieve, sifting the fragments so that those of one size are dropped in one place, those of the next size in an adjoining tract, and so on. The reader can notice this sifting action on a small scale in any little pool into which a stream runs, even in the pools which form on the roads in rainy weather. Each runnel of water carries little pebbles, grains of sand and particles of mud, and when the runnel enters the pool the pebbles are arranged nearest the mouth of the runnel, the sand grains further away, and the small particles of mud furthest away from the side of the pool. To complete the comparison between this deposit and that which takes place in the sea, we may find bits of twigs, leaves of trees, and perhaps some snail-shells washed into the pool, and embedded in the deposits to represent *fossils,* or the remains of once living things which are entombed in the rocks formed in the past.

In Fig. 13 an attempt is made to shew the mode of arrangement of the materials, as would be seen if an upright cutting or section were made through them.

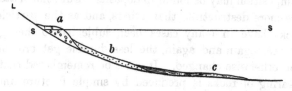

Fig. 13.   Deposits of a coast-line.
a Pebbles.   *b* Sand.   *c* Mud.   L Land.   S S Sea-level.

It must be understood, however, that the sorting is not complete. Some grains of sand will be entangled among the pebbles, and occasionally a small pebble may be carried out to lie among the sands, while a mixture of fine sand and mud will probably form a belt between the pure sand and mud. The distances to which the materials are carried varies of course with the strength of the currents. The pebbles are usually laid down quite close to the shore: the sand forms sand-banks, often dry at low water, extending to no great distance from the coast. The bulk of the mud is no doubt laid down in a strip extending only a few miles away from the coast, but the finer particles are sometimes carried by currents of greater power than usual for a distance of more than one hundred miles away from the shore lines. These deposits, formed of the materials carried in the water in a visible state are forming fringes around the shore lines: they are *coastal* deposits.

Some of the reasons for the formation of definite beds among these deposits separated from each other by bedding-planes were given in chapter III.

We may now say a few words about the composition of the fragments of these rocks. Any fragment broken from another rock may contribute to the formation of new deposits, and accordingly materials of very variable composition may be found in deposits. But some materials are more destructible than others, and as the particles of rocks have in many cases been subjected to the agents of wear again and again, the less durable get broken up, and otherwise changed. It must be remembered that the wearing of rocks is produced by simple fracture and by chemical change, and the substances which withstand fracture and chemical action most successfully will in the long run survive. The hardest rocks, and those with fewest divisional planes, yield fragments which if not acted upon chemically will furnish the hardest pebbles. Accordingly substances like *quartz* and *flint* are often found as pebbles. Quartz or rock-crystal as it is termed, when found in transparent crystals is composed of a substance known as silica, one of the most durable of materials, both on account of its hardness and its resistance to chemical action. Silica is soluble, as we shall have occasion to note later, but very slightly so. Flint is very much the same as quartz. Quartz grains again are the commonest constituent of sands. Other minerals do occur in sands, but as a rule they are far outnumbered by the quartz grains. There is a mineral known as *felspar* which is common in many igneous rocks. It is partly soluble in rain water, and the insoluble part forms what is known as china-clay. This clay is in a very fine state of division, and when the particles are washed away, they may be carried much further than the sand-grains. This clay is not likely to be deposited on a large scale in a state of purity. The smallest chips off other fragments, which may vary greatly in composition, tend to be mixed with pure clay, and the finest deposits or *muds* vary in composition according to the character of these little chips, which may themselves undergo further change after deposit owing to the action of sea water upon them.

The principal sediments formed in water areas by the settlement of fragments are therefore pebble-beds, which when the pebbles are cemented together form *conglomerates*, sands which may be cemented into *sandstones*, and muds often spoken of as clays, which are compacted into *mudstones*, or if laminated are termed *shales*.

We must now pay attention to the materials which are brought into the sea in solution, some of which are brought back into the solid state. It was seen in chapter XI that in inland lakes having no outlet to the ocean some of this material may be precipitated on to the lake floor owing to evaporation. This cannot take place in the ocean, save very exceptionally and on a small scale, owing to the waters of the oceans being in quantity far more than sufficient to keep in solution all the dissolved material which they contain and which is carried into them by rivers. Certain plants and animals living in the sea are capable of extracting some of the substances dissolved in the water, and using them to build up their hard parts, such as shells and bones. After the living creatures die these hard parts may be collected on the sea-floors in a state of sufficient purity to give rise to what are known as organic deposits, and it is these deposits with which we are concerned.

The bones of animals are largely composed of a substance known as *phosphate of lime*: this does to some extent enter into the composition of certain organic deposits but it is not important as regards frequency of occurrence.

The principal materials forming organic rocks are carbonate of lime and silica, of which the former is by far the commoner, giving rise to limestones.

Certain plants as well as animals can extract carbonate of lime from sea water, but the bulk of limestone deposits are formed by the remains of animals. Among those which contribute most largely are molluscs, crustaceans (that is animals of the crab and lobster group), sea-urchins and their allies, corals, and a group of very lowly animals known as foraminifers.

Of course, where animals live in considerable numbers in the strips of sea surrounding the coasts their remains become mingled with the ordinary particles deposited on the sea floor, giving rise to deposits which are partly organic and partly inorganic. Thus we find shelly sandstones and shelly muds, and even shelly conglomerates.

Limestones are known as calcareous rocks, and these mixtures of carbonate of lime and other materials are spoken of as calcareous conglomerates, sands, and muds. The purer organic deposits are usually formed at some distance from the coasts, though, when the sea-waters are devoid of sand and mud owing to special causes, fairly pure organic deposits are built up near the shores.

These deposits are partly due to living beings which exist either fixed to or able to move over the sea-floor, and partly to those which live freely in the waters, especially near the surface; the latter after death fall through the water to settle on the sea-floor.

We first take account of those beds which are formed of the beings living at the bottom of the sea. Certain plants live freely in fairly shallow water, and give rise to limestones in freshwater lakes as well as in the sea.

Of beds formed by animals we may first of all notice the shell-banks. Oysters and other molluscs often form layers of limestone of comparative purity. In the more ancient rocks a certain shell-fish, which though belonging to a different group from that which contains the cockle, whelk, and nautilus, forms shells somewhat like those made by the cockle, often contributes largely to the deposits. These shell-fish are known as brachiopods or lamp-shells. Sea-urchins sometimes occur in numbers in limestones, and in ancient times very important deposits were formed in what is now the British area and elsewhere, by their relatives the crinoids or sea-lilies, as they are sometimes called, though the term lily is fanciful, for they are animals and not plants.

In the clear, shallow waters far away from ordinary coasts the corals build great sheets of limestone below high-water-mark. These creatures only live in sufficient numbers to form their limestone *reefs* in warm seas, a point which must be borne in mind, for the occurrence of coral-reef limestones in our own country in certain rocks may throw some light upon the climatic conditions which have prevailed in past times.

Now we pass on to consider those deposits which are made up of the remains of living beings which swim or float freely in the ocean-waters.

Of these the most widely distributed is a whitish deposit formed very largely of the hard parts of the animals which we have spoken of as foraminifers. Many of these creatures form a *test* (or shell-like covering), often of very lovely pattern, though usually so small that the beauty of the object can only be seen when viewed through a magnifying-glass. The ordinary size of these little objects is about that of a pin's head. They live in countless swarms in the upper waters of many parts of the ocean, and as they die their tests are rained down on to the sea-floor. Where ordinary sediments are being deposited they are mixed with the sand and mud, but as we pass from the shore lines to the open ocean, the mud gets smaller and smaller in quantity, and the proportion of these tests to the mud consequently larger, until at last the ordinary mud forms a very small portion (sometimes only about 5 per cent.) of the whole, the rest being a limy deposit formed of whole and broken tests often mixed with a certain amount of lime derived from minute plants. This deposit is spoken of as *ooze*, the particular material under consideration being a calcareous ooze. It always contains remains of other animals besides foraminifera, often in some quantity. The smallest depth at which it occurs varies according to the nature of the sea-floor, where ordinary sediment is carried no further seaward; but the belt where it occurs may be regarded as lying at depths varying from one mile to over two miles vertically below the surface of the oceans. Much deeper than this it does not extend, owing to other causes which will be presently mentioned. Millions of square miles of the ocean-floors in tropical and temperate regions are at the present day being covered by deposits of this ooze.

Another group of animals known as *radiolarians* are near relatives of the foraminifer, but the tests which they form are not made of carbonate of lime but of silica. Accordingly they give rise to siliceous oozes in much the same way as that in which the foraminifers form calcareous ooze, but they extend to a greater depth than the latter, having been found at a depth of five miles. Other living things, both animals and plants, can secrete silica; among the animals are sponges, many of which form a sort of skeleton or spiny thread-like structure.

The organisms which extract silica from the sea-water also usually add to the materials formed in the calcareous oozes, an addition of some importance as throwing light on certain siliceous materials found among calcareous rocks of past time.

At depths of over 5000 yards the calcareous tests become dissolved, which accounts for the calcareous oozes having a more or less definite downward limit. Below that depth vast tracts of the ocean-floor are covered with a reddish or greyish clay of extreme fineness, mainly composed of volcanic materials, but containing also little globules of metal, which are minute meteorites, having reached the earth from some remote source. The deep-sea clays must be formed with extreme slowness.

CHAPTER XIII.

ON MOVEMENTS OF THE EARTH'S CRUST.

We have now considered those changes on the surface of the land whereby the rocks of the earth's crust are broken up into fragments or removed in solution, and the materials formed by the processes of waste carried seaward.

From what has been said in previous chapters it is clear that if these changes went on unchecked for a sufficient length of time the whole of the land surfaces would be reduced below sea-level, and the materials derived from the waste of the land arranged as sediments upon the ocean-floor. There they would lie undisturbed unless brought above the ocean surface in some way.

We can readily find out that parts of the ocean-floor covered by these sea-deposits have been converted into dry land, for in many parts of the world, including large tracts of our own country, we find that the rocks seen in cliffs, stream-sides, and quarries contain the remains of such creatures as corals, sea-urchins, sea-lilies, crabs and the like, which only live in the sea, and their abundance and modes of occurrence forbid the idea that the individual relics of these sea creatures have been brought on to the land one by one, as sometimes happens. They have clearly been embedded where they now lie, indeed we often find them attached to the old sea-floor.

There are two ways in which a portion of the sea-floor would become dry land. If in any way a bulk of the sea-water could be moved from the sea, or from part of the sea, the shallow parts surrounding the coast would be turned into land-surfaces. The same thing would occur if a part of the earth's crust were raised up, so as to emerge above sea-level, or if the bottom of the sea-floor sank so as to let the waters down, when they would recede from the coast-lines.

It is not now denied that some appearances of sea-beds above the ocean surface may be due to movement of the waters apart from any movement of the earth's crust above or below the sea-level, but there is much direct evidence that the earth's crust undergoes movement.

Changes of level of parts of the land surfaces have actually been observed during earthquakes, indeed the quaking is often the direct result of such changes of level. To take one instance, already mentioned, in 1855 an earthquake occurred in New Zealand, and it is computed that a tract of country near Wellington between four and five thousand square miles in area was raised to an extent varying from a foot to nine feet above its former level.

Striking as are the effects of earthquakes it may be doubted whether those movements which are accompanied by violent shakings of the surface produce so much effect as the slow changes marked by no sensible disturbance which may go on continuously for long ages.

Be this as it may, and whatever the nature of the movements, their effects are readily ascertainable by examination of the stratified rocks of some districts.

It was pointed out in chapter III that the beds which occur on the land-surfaces are seen to be thrown into great folds, so that they may dip at any angle from a horizontal plane, and may be actually on end, or even overturned. Again, in chapter IV we noticed that parts of these beds are often displaced by means of faults. The beds were not formed in folds; they were deposited in flat layers, and have been folded afterwards. That shews that beds have been moved, but as these beds compose parts of the earth's crust, that is merely another way of stating that those parts of the earth's crust have been moved. An examination of beds therefore proves that movement of the actual crust of the earth has taken place in past times.

The appearance of the dry land deposits formed below the sea implies elevation, using that term with reference to the raising of blocks of the earth's crust above sea-level, or to a greater height above sea-level than that at which they previously stood.

We also find proofs that tracts which were once dry land were *depressed* below sea-level, though the proofs in this case are not so easy to find as in the case of movements of elevation.

At Whitehaven, in Cumberland, beds of coal are found beneath the sea. Now these coal-seams shew signs of having been formed above the sea-level, and their present existence below that level points to depression. Of later date than these coal-seams are old soils upon which trees grew which are found along many parts of our coast below the level of the sea.

Again, the sea-lochs of Scotland and the fjords of Norway have not been carved out, as sometimes stated, by the waves of the Atlantic. In the narrow Norwegian fjords bounded by high cliffs the sea is often calm when storms are lashing the ocean outside into high billows. These indentations shew by their structure that they are valleys formed by river erosion aided in some cases by ice ; for their structure below water agrees with that above. The waters of the ocean have in part drowned the lower parts of valleys, which were formed when the land was at a higher level.   It has since been depressed.

If the reader will turn back to the tenth chapter, and read the account of unconformities, he will see that the existence of an important unconformity shews that there has been elevation, during which erosion of the earlier marine deposits took place, followed by depression, during which fresh deposits were laid down upon the upturned and eroded edges of the former ones.

Such are some of the reasons we have for assuming that parts of the earth's crust have been moved up and down in past time, and are being so moved at the present day.

By these movements then we can account for the former conversion of land tracts into sea and of sea-floors into land, the frequent occurrence of which is one of the most important kinds of events with which the geologist has to deal.

If the reader will take a stick of sealing-wax and try to bend it quickly it will snap in two. If, however, he tries to bend it very slowly he may be able to make it into the shape of a bow, but if after he has got it into this shape he tries once more to bend it quickly it will snap. Similarly, rocks may be curved into folds, such as we noticed in chapter II, if bent slowly, whereas, if bent more quickly they will snap, and faulting may take place along the lines of breakage. Accordingly, the rocks of any area may be affected by folds without faults, or by faults without folds, or by folds accompanied by faults.

The degree to which rocks may be folded depends not

only on the time taken over the process, but also on the conditions under which the rocks exist. There is reason to conclude that when rocks are deep down below the surface and weighted by thousands of feet of rock resting upon them, they will fold more readily than when near the surface.

Again, if we take a stick of sealing-wax, in order to fold it we must press the two ends towards one another. If we try to pull them apart we cannot so fold the stick, but may break it. Folding then is the result of compression of the rocks sideways, while breakage may be brought about by stretching.

We have proof that in many areas the rocks have been thrown into folds by side pressure, while in others they have been stretched. In the former the rocks are folded, the degree of folding depending largely on the amount of side pressure; in the latter the rocks are faulted without being folded, except in a minor degree. The former areas are often occupied by *mountain ranges*, while the latter tend to give rise to *plateaux*, or elevated surfaces of a more or less level character. To this matter we shall allude in the next chapter.

Geologists have naturally been led to consider the causes to which movement of the earth's crust is due. A number of circumstances exert a control over this movement, and the whole subject is one of great complexity. We can here only refer to the principal cause which is supposed to set up crust-movements.

Those who have read Sir Robert Ball's *Primer of Astronomy* will have learnt that the earth was once a hot, molten mass of matter. It has been losing its heat by degrees. Now, through long ages, the outer crust has been solid, and its temperature, as shewn by the nature of the rocks of long past ages, cannot have been very different from what it now is, for in those remote times creatures very like those now living inhabited the outer waters. But the earth's interior is still hot, as we know from many facts, the most striking of which is the bringing up of molten rock from the interior to be poured out on the surface by volcanoes. The hot inside is therefore now cooling, while the outer crust has a fairly constant temperature apart from changes due to climate. The heat of the inside is conducted through the rocks of the crust, and given off into space.

Now, most substances, as we know them, shrink when cooling, and not only so, but they shrink more for a given loss of heat when very hot than they do when not so hot. It is therefore fair to argue that the inside of the earth shrinks more than the outer parts, and must tend to get smaller. The diameter of the earth is about 8000 miles. Let us imagine that a crust five miles thick was practically not shrinking, while the inner mass was shrinking. Such a mass having at one time a diameter of 7990 miles would after a long period have a smaller diameter, say 7980. If nothing else happened, a space five miles deep would exist between the central core and the outer crust. But the weight of the crust would not permit this space to exist; the rocks of the crust would settle down in wrinkled masses on the shrunken core, and if the outer surface of the crust were first smooth, it would now be marked by uneven outlines. The ridges might form continents and mountains, and the hollows sea-beds.

The comparison of the earth to a withering apple, which has often been made, illustrates these possible changes. In the case of the apple the withering is due to loss of moisture and not to loss of heat. As the main mass of the apple contains more water than the rind, it tends to shrink more when the moisture is given off, and the outside of the central part becomes smaller than the inside of the rind. Accordingly the rind becomes wrinkled.

But, if we grant that contraction of the earth's interior is the main cause of crust movement, we must add that it may set up many minor actions which affect the nature of the movements. For instance, molten rock may be squeezed from one place to another in the earth's interior, and the crust may sag down at the place beneath which the molten rock was squeezed away, and rise in dome-like swellings in the parts to which the rock is squeezed. By the loss of the heat of the earth's interior then we find a means of making up for the destruction of continents by the agents of erosion.

Erosion destroys continents, the wreckage is sorted afresh beneath the waters of the oceans, and this resorted material is made into fresh land areas by earth movement. The land is the area of destruction by erosion, the sea is the area of construction by deposit, and underground change is responsible for reproduction by earth move-ment.

These three kinds of change then bring about some of the most important events with which the geologist has to deal when studying the earth apart from the life which exists or has existed upon it.

It is interesting to see that whereas, as noted in chapter X, the energy furnished by the sun's heat is responsible for destruction of land masses, that due to the heat of the earth's interior is utilised in formation of fresh masses of land.

CHAPTER XIV.

INFLUENCE OF EARTH MOVEMENT AND EROSION ON SURFACE FEATURES.

ONE of the most fascinating of the pursuits of the geologist is the attempt to explain the origin of the form of the earth's surface.

The main features of the land tracts are due to the combined effects of crust movement and erosion, save where modifications are produced by accumulation of materials on the land, and with the exception of volcanic mountains, which will be considered in a later chapter, these are somewhat unimportant. Below the waters of the ocean, on the contrary, the effects of earth movement are mainly modified by accumulation.

Whatever be the exact shape which the earth would have assumed, apart from crust movements and erosion, there is little doubt that on the whole the surface would closely approach that of a smooth sphere. The large earth movements are probably responsible for the actual existence of some tracts as land and others as sea. We need not however consider the features of the oceans, which are not visible to us, and which indeed are even yet not largely known. It is the outlines of the land, with their varied details of mountain, vale, plateau and plain, which we are naturally led to consider.

That the present land features are to some extent due to movements of the earth's crust is obvious from what we have already written, for had these movements not occurred our continents would no longer exist; they would have been worn down by the agents of erosion to a level some little depth below the surface of the ocean waters.

Were the surface features entirely due to crust movement without being modified by erosion the nature of the surface features would depend upon that of the results of the movement. Notwithstanding the effects of erosion a direct relationship is frequently traceable between the effects of movement and the present surface features. It will be convenient therefore if we consider at the outset the types of scenery which would be produced as the effects of movement without erosion.

Fig. 14. Beds thrown into folds giving rise to Mountain-chains, and intervening hollows.

A, Even-sided fold.   B, Uneven-sided fold.

If the reader will place a book on the centre of a table covered by a cloth, and push the cloth towards the book, parallel rows of folds will be set up in the cloth, consisting of ridges and troughs. Similarly, if the earth's crust be laterally compressed, ridges and troughs may be formed, the height or depth of each being possibly greater than its width, possibly less, but not being very small when compared with the width. Such a series of folds would give rise to mountain chains, with trough-like hollows between, and the surface of the mountain chains would be convex, those of the hollows concave. Furthermore, the beds which composed the mountain chains and hollows would be arranged with their bedding planes parallel to the surface of the ground, as shewn in Fig. 14. The folds may have their two sides similar, as at *A*, or one side may be steeper than the other as at *B*.

In many cases smaller folds, which are often overturned, occur as wrinkles on the sides of one great fold, as shewn in section in Fig. 15. As the result of this structure, which is known as *fan-structure*, the arrangement of the beds in mountain regions is often very complicated.

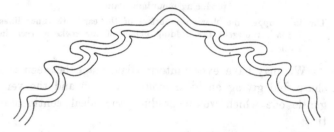

Fig. 15.   Fan-structure.   (Two strata only are shewn.)
The upper surface as it would appear if no erosion took place.

It need scarcely be remarked that in no place do we find a mountain range shewing such a structure with the outline due to movement alone. If ordinary erosion did not take place landslips would certainly remove the overhanging portions.

The tops of the ridges and the bottoms of the hollows need not be absolutely level. In this case parts of the hollows may be occupied by standing water, giving rise to one kind of lake.

Suppose instead of a sharp fold we have a very gentle arch, the height of the arch being very small as compared with its width. If the width is very great a large tract of land may thus rise out of the sea, and its slopes would be very gentle. Such a tract might however have sharp wrinkles here and there, as shewn in section in Fig. 16. The continental area with generally small slopes would then be diversified by mountain ranges.

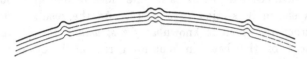

Fig. 16.   Gentle fold of continental type with smaller wrinkles
producing mountain-chains.

The thick upper line shews the surface of the earth, the finer lines
    beneath represent the bedding planes of the rocks as seen in
    section.

We may have every intermediate stage between the sharp fold giving an ideal mountain chain and the very gentle one which would produce the ideal continental tract.

We have seen that in areas where the earth's crust undergoes stretching, fracture will occur, and parts of the crust will settle to a lower level than others. If erosion did not modify the outlines of the surface we should have as a result of such crust movement some such arrangement of beds and appearance of the surface features as is shewn in section in Fig. 17. When this structure occurs on a large scale the uplifts form one kind of *plateau*, with intervening depressions; if the scale be smaller the uplifts produce *block-mountains*. In the case of movements of this sort also, parts of the depressions may be occupied by lakes.

Our actual continental surfaces, even apart from erosion, would as a rule present very simple outlines, for in a large continental area, different parts have been affected by different kinds of movement.

Fig. 17.   Block-mountains with intervening hollows.   The surface shewn as it would appear if no erosion took place.

ff... faults.   Two beds *bb* are represented lying parallel with the surface.

Also, apart from the actual influence of erosion on the surface outlines, its result is often to wear away beds, of which the bedding planes would have been parallel to the uplifted surface, and to expose beds which lay unconformably beneath the upper ones; the bedding planes therefore abut against the actual existing surface.

From what has been said in chapters v—x, it is evident that the agents of erosion produce very important modifications in the surface features of the lands. These agents can only operate as the result of uplift, and the greater the uplift, the greater on the whole is the activity of the agents. If we imagine a tract of former sea-floor to become raised, erosion commences as soon as a part of the old floor emerges above sea-level, and continues while elevation takes place. A land tract is therefore due to the greater power of uplift to form the tract than of the agents of erosion to destroy it. If the uplift ceases the agents of erosion still go on working and will at last, as has been already stated, reduce the tract to a level.

The results of the agents of erosion on the surface features depend therefore on three things: (i) the nature of the rocks; (ii) the character of the erosive agents; (iii) the length of time during which erosion has been going on, and whether uplift has ceased in the district which is undergoing erosion.

We will first consider the effects of the nature of the rocks. It is abundantly clear from what has already been said that some rocks resist erosion more than others. This is the case during the processes of weathering and also during the occurrence of the rasping action of wind, rivers and glaciers. A number of characters of the rocks affect the results, different characters being of different degrees of importance in the case of various agents. Hardness, nature of divisional planes, and degree of solubility of the rocks, all produce their effect, but on the whole the hard rocks tend to stand out as eminences, the soft rocks to be worn away.

The principal variations in the type of erosion produced by the different agents have already been considered in detail, and it here only remains for us to put forward some general statements. Apart from weathering, there is no doubt that rivers play the most important part as agents of sculpture on the land.

It has been seen that as long as a river has any energy available for the process of deepening its bed the valley will continue to be cut deeper. Accordingly, if we commence with an ideal surface such as we have supposed to be formed by crust movement alone, that surface will become notched by the action of rivers. With a mountain ridge, as the rivers do not flow down the ridge over its whole extent, the parts left between two streams will stand out as minor ridges, running on the whole at right angles to the main ridge. As the rivers work back at their heads, the main ridge will be notched in places, and thus will be cut up into a mass of land with minor ridges, with the main ridge consisting of individual mountain summits with notches or passes between the adjoining ones.

The tendency of rivers is to carve the lines of the valley bottoms into concave curves, and as the rivers on each side of the main ridge do this, a convex mountain uplift like those shewn in Figure 14 will by degrees be converted into a tract with two concave curves, as shewn by the dotted line in Figure 18, where the unbroken line represents in section the curve which would be formed by uplift alone.

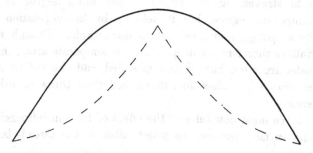

Fig. 18.   Curves of stream erosion modifying slopes of mountain-chain. The dotted lines are the curves of stream erosion.

On gentler slopes the valleys will also be deepened so long as the rivers cut downward, and a plateau-like surface will be cut up into fairly flat plateau-tracts with valleys formed by the rivers.

In high latitudes or altitudes, as in the arctic regions and the higher parts of the Alps, running water cannot affect the high peaks to any extent. Their sides are largely determined by the action of frost in splitting off fragments along the more important divisional planes, and accordingly these mountains often possess very straight sides. Such mountains are said to shew house-roof structure.

A similar structure is often found in the mountains of desert regions where stream action is again of little importance, the action being similar to that produced by frost in cold regions, but instead of frost, alternate contraction and expansion, owing to the cold and heat, cause the splitting action.

In tropical regions the great rainfall causes luxuriant vegetation to grow in many places. This to some extent checks the action of streams, as the water is not formed into streams on the slopes to the same degree as in temperate regions, but is held up by the vegetation as by a sponge, and slowly soaks downwards. Though the valleys therefore are deepened as in temperate areas, their sides are often but slightly modified, and the valleys are separated by ridges less diversified than those of other areas.

We must now refer to the effect of time in influencing the surface features, to which allusion has already been made.

Rivers, wind, and glaciers alike can only cut downward when the slope of the ground on which they are working is sufficient to produce enough energy for downward erosion. As long as uplift goes on sufficiently rapidly this downward erosion will tend to cut up the surface which would otherwise exist as the effect of uplift only, so as to produce a greater difference between height of hill and ridge summit and of valley bottom or lower part of floor of wind-swept tract. At first the actual features due to uplift may not be greatly modified; the convex mountain ridge may retain its convexity, the gentle slope its appearance of a plain, and the block mountain or plateau its flat top and steep sides defined by the fault fissures.

When uplift ceases, or becomes so slow that the intervening ridges are lowered more quickly than their valley bottoms, the diversity of features due to downward erosion will become lessened, and the process will go on until the whole land can be lowered no further. The hard rocks will still resist for some time, but if the soft rocks can be cut down no further the day must eventually arrive when the hard rocks also will be worn away, and the country will be almost a plain, unless in the meantime the sea has done its work and reduced the whole region to a plain below sea-level.

CHAPTER XV.

VOLCANOES AND IGNEOUS ROCK-MASSES.

It was pointed out in the second chapter that rocks could be divided into two great classes, those of one class being of aqueous origin, while those of the other were spoken of as igneous rocks.

We have considered the general character and modes of production of the former class and must now pay attention to the igneous rocks.

When we come to study the igneous rocks of past times we find that some of them have obviously been formed as the result of the existence of volcanoes, while others have become solid at considerable depths below the earth's surface, and these need not have been connected with volcanoes, though sometimes volcanoes have been formed above the places where they existed.

Accordingly, igneous rocks have been divided into two groups; those of the first group have formed parts of volcanoes, and are therefore named *volcanic* rocks; those of the second have been solidified at considerable depths in the earth's interior, and are called *plutonic* rocks after the god Pluto, who was supposed to inhabit the lower regions. There are it is true some rocks which are links between the volcanic and plutonic rocks, and we may speak of these as *intermediate* rocks.

As the volcanic rocks may be studied in the act of formation at the present day, while the others are being formed out of sight, it will be convenient to consider the volcanic rocks first. To do this we must take into account the mode of production of volcanoes and their general structure.

It is not very easy to give a short and at the same time accurate definition of a volcano, but a brief description will serve in place of a definition.

A volcano may be regarded as consisting of an opening communicating with the interior of the earth, through which rock, which has on the whole been liquid shortly before its appearance at the surface, is brought to the surface either in a molten condition as *lava*, or in fragments derived by explosive action from the recently solidified mass.

The materials which are brought to the surface accumulate around the opening, and in many cases build up a hill, which is the object to which the name volcano is popularly applied, though the existence of a hill is not necessary, indeed some of the most remarkable volcanic accumulations which occur on the earth's surface have been brought up in such volume that they give rise to plateaux and not to mere isolated hills.

We will in the first place describe the mode of formation of a simple volcanic hill, in order to illustrate the general nature of the materials which are brought to the surface. This description will serve to explain also the main characters of the volcanic rocks as a whole. Afterwards we can point out such variations in the modes of building up of the materials as are of special interest.

It has already been seen that as the result of crust-movements, and particularly of those movements which result in stretching of the rocks, cracks are formed. Along these cracks certain matter may be brought up to the surface. If the volcanoes are formed as isolated hills the substances will not be brought up along the whole line of the crack, but at definite points which are spots of special weakness, along which the matter will be forced more readily than elsewhere. Owing to this cause we usually find volcanic hills arranged in lines, these lines being along the cracks, as shewn in Fig. 19. Taking

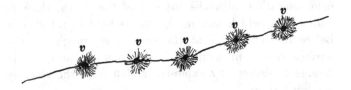

Fig. 19. Plan of a crack with volcanoes *v v*... formed along it at points of weakness.

into account one of the spots of weakness the first process is the enlargement of the crack at that spot. This will probably be due to explosive action, the sides of the crack being blown out, so as to give rise to an opening, probably with an outline approaching the circular form, it being the top of a shaft going down to the rocks which will afterwards furnish the material for building the volcanic hill.

The rocks which have been blown out will tend to form a little ring around the opening. This ring is the beginning of the hill, and the rocks of which it is formed will vary according to the nature of those bordering the shaft or pipe.

The pipe may now form a free means of passage for

the rock which exists in a molten condition at its lower end, and in the state of affairs now brought about, some of this rock can be brought to the surface, if there is any agent capable of raising it. It may be raised in more than one way, but actual observation of a number of existing volcanoes shews that one of the agents which raises the rock is water when converted into steam. If the top of the molten rock becomes solid, it acts as a cork to the still liquid matter beneath. Now, if we cork a bottle containing water, and boil that water, either the bottle bursts, or the cork is blown out. In the case of the volcano, when the water in the molten rock below is converted into steam, the cork-like plug of solid rock above is blown out, but naturally breaks into pieces during the process, and we may therefore notice, if fortunate enough to see one of these explosive outbursts from a safe place, great jets of steam coming from the orifice, accompanied by a cloud of broken fragments of various sizes, from huge angular stones to fine dust.

We must here say a word about the reason why the water becomes steam, or, rather, why it exists as water in molten rock which is heated far above the point at which water in a kettle would change into steam.

It is not easy to make good tea on a high mountain because the water up there boils before it has been raised to the temperature necessary in order to make it boil at sea-level. This is owing to the pressure of the air, which is less on the mountain-top than at the sea-level. Now, if the pressure of the air makes this difference, we should naturally expect that the pressure of the masses of solid rock which lie over the molten rock in the inside of the earth would produce a greater difference, and there is no doubt that the water which we know exists in molten rock will remain as water until the pressure is removed. That takes place below the opening of the volcanic pipe when the pipe is emptied of solid rock to a sufficient extent, and then the water is converted into steam and blows out the remaining solid rock.

The first blowing out of the cork-like mass of rock will cause the ring formed by the rock from the side of the pipe to receive an addition, consisting in this case of true volcanic rock, and our baby volcano thus grows in height.

Explosion will succeed explosion, each adding to the materials of the hill. On the outside of the rim the loose fragments will be piled up in a slope which is that of the angle of rest of the fragments, just as the materials of a gravel-heap get piled up. But there is a difference between the gravel-heap and the volcano, for there is no hole below the gravel-heap. Now, in the case of the volcano some of the material falls back into the opening, and accordingly there is another slope towards the opening at the angle of rest of the blocks. If the opening be circular, the hollow thus formed will be like a funnel of which the pipe of the volcano is the stem, and the funnel-shaped opening above the pipe is known as the crater of the volcano. As the material gets piled up the bottom of the crater is gradually raised above the old surface of the ground on which the volcano commenced to grow. At this stage of the history of the volcano, a cross-section will appear somewhat as in Fig. 20. The fragments will be arranged in layers like those of aqueous rocks, owing to pauses between different explosions, and the layers will slope inward below the crater, and outward away from it, as shewn in the figure. Each layer will thin away from its thickest point below the rim of the crater, toward the outer ring of the hill and also toward the orifice.

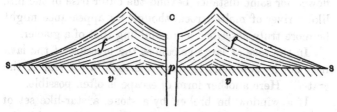

Fig. 20.   Section across a young volcano.

v v ordinary rock.   C crater.
S S former surface of ground.   *f f* fragments of rock in layers
p   pipe.   forming volcanic hill.

Such is a simple type of volcano, and in its making no lava has been poured out upon the surface. It is not to be supposed that a volcano must necessarily begin in this way. Lava may be and often is poured out before any fragments of volcanic rocks are shot out. We have only taken the commonest method of origin of a volcanic hill for convenience.

After a hill has been built up entirely of ashes the volcano may become *extinct*: no further eruption will occur, and the section of the volcano will retain the appearance presented in the figure.

We have now to turn our attention to the bringing up and pouring out of lava on the surface.

This lava has to be pumped up in some way, and as it rises it exerts a considerable pressure on the sides of the pipe. It may fill the crater, and flow over, or it may more probably burst the side of the crater and pour out through the breach. This has occurred in the case of many of the now extinct volcanoes in the district of Auvergne in Central France. The hills were first built up of fragments in the way described, molten rock then rose, burst one side of the hill around the crater, and often flowed for some distance beyond the outer base of the hill, like a river of molten rock, though its appearance might be more truly described as resembling that of a glacier.

In the case of large volcanoes the pressure of the lava may not be sufficient to form a breach in the side of the crater. Here another form of escape is often possible.

If a window be broken by a stone, a star-like set of cracks spreads out from the point where the stone hit the glass. Now, an explosion in the pipe of the volcano may cause the rocks of the volcano (which will have been made solid by chemical change and pressure of rock above) to crack, and the cracks will here also form a star-like pattern, starting from the pipe, but as the rocks of the volcano are of considerable thickness, these cracks, unlike those of the glass, will be through a thick mass of rock, and each crack will tend to be upright. Along these cracks molten matter from the pipe may be forced and some of this molten matter will probably reach the surface of the volcano on its side, and come out at a point some distance below the rim of the crater. It will then flow down the side, becoming solid by degrees, until at last it ceases flowing.

The rock in the crack also becomes solid, and the former crack instead of being a plane of weakness, is now one of extra strength. The next outburst of molten rock will therefore probably be through another crack, and so the lava-flows are poured out, now from one point, now from another, until in course of time every part of the volcano's surface may have received its share of once molten rock and the hill will have a shape as regular as though it had been built of fragments only.

The cracks filled with once molten rocks are spoken of
as *dykes*; as we shall see later, such dykes are not limited
to volcanoes, but may occur along cracks extending into
the ordinary rocks forming the earth's crust.

A section through a volcano composed of lavas and
fragmental rocks is shewn in Fig. 21, where the dykes are
not represented, as being usually upright, and in the
direction of the section, they would not as a general rule
be cut through.

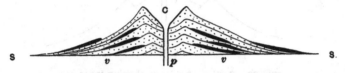

Fig. 21.   Section across a volcano formed of lavas and fragmental rocks.
The lavas are black, the fragmental rocks dotted.

v v ordinary rock.	*p* pipe.
S S former surface of ground.	C crater.

We will now consider some of the changes which may
occur in a simple volcano.

Small cones known as parasitic cones are often built
up on the sides when a new orifice is formed. As they
are generally small compared with the main hill they do
not often render its outline irregular.

Sometimes the top of the volcano is blown off by an
unusually violent explosion, or it may sink down into the
molten rock below. After this a new cone may be built
in the hollow thus formed, as shewn in section in Fig. 22.

Again, two volcanoes may grow into one, giving rise to
a double-peaked hill, each having its own crater; or we
may even find a group of such hills joined together.

We must now pay some attention to the characters of
the volcanic rocks which form these hills, in order to be

able to recognise rocks formed in the ways above described when we meet with them among the older rocks of the earth's crust. We have said something about their nature in chapter II., but we are now in a position to learn more about them. We will firstly consider the fragmental rocks which were blown out in the solid state, and then the lavas.

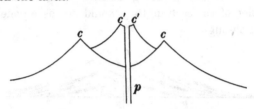

Fig. 22. Section of volcano shewing inner cone.

c c rim of old cone.   *c′c′* rim of modern cone.   *p* pipe.

The fragmental rocks are spoken of as volcanic ashes, though it is not to be supposed from this name that they resemble the ashes of a grate. The latter are the remains of materials which have been *burned*. No burning of the volcanic ashes has taken place, and indeed the term "burning mountain" sometimes given to a volcano is misleading. The volcanic ashes are simply broken pieces of once molten rock, the breakage having taken place in the way described. The cindery appearance which the fragments sometimes present is due to the way in which the molten rock of which they are formed has become solid, as will be presently explained.

Ashes may be distinguished from ordinary deposits by their composition rather than by their structure. In trying to discover if a rock of ancient date is a volcanic ash it is well to look for the larger fragments contained therein. If they are angular and have the composition

and structure of igneous rocks it is probable that the
containing rock is a volcanic ash. But many other pieces
of evidence, into an account of which we cannot now enter,
must be regarded before the geologist may be satisfied
that he is dealing with an ash.

The ashes which are showered upon the slopes of
volcanoes situated upon the land rarely include traces of
living things save occasionally the charred stumps of trees,
but when the ash falls on the sea-floor it may and often
does enclose the remains of the organisms which inhabited
that sea, and accordingly among the rocks of the past we
occasionally find ancient ashes enclosing the relics of
marine creatures.

The structures by which we may distinguish lavas
from sediments are often very apparent. It has already
been said that igneous rocks are crystalline or glassy. If
glassy, their formation by cooling of molten rock is certain,
and there is often little difficulty in detecting the igneous
character of a crystalline rock, for the crystals which it
contains differ on the whole from the commoner fragments of sediments, even when their outlines are not well
marked, as is generally the case. It is rare to find a
sandstone composed of the minerals which form igneous
rocks, for in the processes of erosion, as already stated,
some of the materials become rotted by solution, and the
broken fragments become sorted. Again, the finer igneous
rocks which may outwardly resemble the fine-grained
sediments when the crystal-grains which form them are
too small to be seen with the naked eye are usually much
harder than the sediments.

A study of recently-formed lavas shews us other
structures which we find useful when trying to settle
the origin of an ancient rock.

Lavas poured out from volcanoes usually become solid at the surface when the inside is still molten. This solid surface often presents peculiar characters. Sometimes the solid crust cracks and jets of still liquid rock are squirted out, and spread out as wavy masses such as one may often see on the top of a mass of cooled pitch. At other times these molten jets take strange rope-like forms, when the surface is spoken of as a *ropy* surface. When the lava is highly charged with drops of very hot water, this water flashing into steam close to the surface causes the crust of the flow to be marked by a number of holes which appear like miniature caverns. It is this which makes the lava surfaces appear as though covered by a mass of cinders, and the lava affected by this structure is known as *cindery* lava. As the molten rock moves onward the cindery surface gets broken up, and the top of a flow which has almost ceased moving may appear like a heap of cinders rolling slowly onward. Cindery fragments are often found in volcanic ashes.

There are also certain structures formed in the interior of some lava-flows which remain for notice. If the reader will look at a dish of impure honey he may note patches of a brown colour irregularly scattered through the straw-coloured mass. Now pour this honey on a sloping surface, and it will flow slowly down like a mass of lava. While so doing, the irregular brown patches will be drawn out into long streaks, giving the honey an appearance of being made of layers of brown and straw-coloured matter. When the flow is partly arrested, these layers may be wrinkled up so as to produce an appearance like that of contorted beds of rock. Similarly, in the case of lava, where clots exist of different structure or composition from that of the rest of the rock, as a result of flow these clots get dragged out into streaks giving a ribboned appearance to the lava, and as in the case of the honey, these ribboned layers may be twisted.

Sometimes the lavas contain large crystals, perhaps an inch in length, which have been brought up from a considerable depth where they formed in the molten mass. The ribboned layers will stream round these crystals as foam in a torrent streams round a stone whose top is above the water level. Indeed this foam on the eddying waters of a pool presents a striking illustration of the appearance of the flow-structure of lava described above, as it would appear if we cut a section through the lava.

One more structure must be noticed. We have seen that a cindery structure is produced by escape of steam at the surface of the lava. Now the pressure above the more deeply seated parts of a lava-flow is often insufficient to keep in the liquid state the highly heated droplets of water which the lava then holds. Each of these water-drops as it flashes into steam forms a hollow, usually about a quarter of an inch across, filled with the water-vapour. This hollow will tend to have a globular form, and if the lava does not move after the water has turned to vapour, the globular form will be retained. Usually the lava moves onward after these hollow globes are formed, and they then change their forms. To illustrate this change let us return to our honey. Cause a few bubbles of air to be enclosed in the honey before it is poured on the sloping surface. These bubbles will have a globular form. As the honey moves down the surface, each little globule is dragged out, so that it becomes longer in the direction of the flow. This is the case also with the steam-holes in the lava. They are dragged out, so that after the lava has ceased flowing, each hole has its long axis parallel to the direction of flow. These hollows, or *vesicles*, are found in many lavas. Pumice is lava which contains an exceptional number of unusually small vesicles.

In course of time, as the lava is subjected to action of water working into it, various substances are dissolved from the lava or some overlying rock, and some of these substances are often deposited on the sides of the vesicle, as the water there evaporates. Lining after lining of this substance may be formed until the vesicle is filled with the mineral. These minerals are very often white, giving the rock with its elongated vesicles the appearance of almond toffee.

It has been stated that the products of volcanic eruptions do not always produce hills. Some of the most extensive eruptions have given rise to plateaux, as in various parts of the arctic regions, in the western territories of North America, in what is now the west of Scotland, in Central France and in the peninsula of India. These plateaux are formed of extensive sheets of lava which have spread far and wide from the place where they reached the surface. They are often nearly or quite flat, and are usually accompanied by little fragmental matter. It has been suggested that the lavas of these plateaux did not break out at definite spots along a crack, but welled up from the crack continuously for a considerable distance and spread out on either side.

We may now pay attention to those rocks which are solidified below the surface, considering in the first place the characters of the rocks, and secondly the way in which they occur.

The more slowly a molten mass is cooled, the coarser as a rule are the crystal grains which enter into its mass.

The plutonic rocks therefore are usually much more coarsely crystalline than are those of volcanic origin. Again, as the pressure under which they become solid does not allow water-drops to flash into steam they do not contain vesicles.

The reader should if possible compare a plutonic rock, of which granite may be taken as a type, with a volcanic rock such as basalt, and if he has access to a museum, it will be well to examine a series of plutonic rocks and another of rocks of volcanic origin.

The rocks which are intermediate between the plutonic and volcanic types are usually less coarsely crystalline than the former and more so than the latter. They are rarely formed of glass, and do not often contain vesicles.

The plutonic rocks and those of intermediate nature are forced in the molten state among other rocks, and occur in various ways. Sometimes we find huge irregular masses occurring on the surface (where they are now exposed owing to erosion) over many miles of country: such are the granites of parts of Ireland, and of Devon and Cornwall. Again, we find mushroom-shaped masses which have been forced along bedding planes. Or the rock may be forced along these planes, over a wide area, with flat tops and bottoms, giving rise to a sheet of rock in some respects resembling the lava of a volcano, but differing inasmuch as it was not poured over the surface of the land. Many masses have been forced along joints, occurring as upright walls among the other rocks. These are *dykes*, and one type has already been considered, which is formed in the cracks of the volcanoes themselves. String-like veins may run irregularly among the rocks. Lastly, the pipe of the volcano may be filled with a cylinder-like plug of once molten rock.

Concerning the causes of volcanic action and the injection of molten rock among other rocks we shall say little. To produce a volcano three things are required: (i) a means of bringing rock into the molten condition, (ii) a line of weakness through which it may come to the surface, (iii) an agent to bring it up.

Various suggestions have been made to account for the conversion of solid into molten rock. Into these we will not enter, but merely remark that it is not necessary that in all cases the process should be the same.

The production of lines of weakness owing to disturbances of the crust has already been considered.

Lastly, we have seen that in some cases the agency of steam is employed in raising lava to the surface. In other cases the mere pressure of the rocks of the crust as they sag down into the molten rock below may cause the latter to well upward through cracks and to spread over the surface.

CHAPTER XVI.

ALTERATION OF ROCKS.

IT is not hard to find out that many rocks have been to a greater or less degree changed after they were formed. It is easier to see that these changes have taken place in aqueous rocks than in those of igneous origin, but we know that changes actually do occur in each class of rock; in the aqueous rocks after they were deposited, and in the igneous rocks after they have become solidified.

Many of these changes are not great enough to disguise the original mode of formation of the rocks, while others have gone on to an extent which has so altered the nature of the rock that it is difficult, and in some cases impossible, to judge how the rock was at the outset formed.

The changes which are most easily studied are those slight ones which have caused the loose particles of sediments or of volcanic ashes to be welded together, so that the rock particles cannot be separated like the loose grains of sand, but the rock must be broken with some force to detach portions from the parent rock.

As a consequence of these changes the loose pebble-beds of the sea-shore are often represented by compact conglomerates among the older rocks; the shifting sands of sandbanks are hardened into sandstones, and the muds of mud-flats into mudstones, while the oozes and other organic deposits are converted into limestones and flinty rocks. This alteration, whereby loose particles are welded together to form *solid rock*, is due to two main causes, namely pressure, and the deposit of material from solution between the grains of the rock.

The effect of pressure in binding together particles which were previously loose can be shewn by experiment: various substances when powdered and then covered by heavy weights will be compacted to form a solid mass. Now, when great thicknesses of sediment have been laid down, the weight of the overlying sediments must affect those beneath; thus, a mass of particles of a mud deposit weighted by some thousand feet of overlying rock, will be compressed, so that the particles adhere to one another, and the rock is changed into a mudstone.

Deposits composed of larger fragments often become solid rock owing to deposit of material from solution. We have already seen that water, especially when it has dissolved certain gases, can also dissolve some of the substances of rocks. These may be again brought into the solid state as the water evaporates. Now, water can find its way between the little crevices which separate the particles of such rocks as shingle-beds and sands, and if that water evaporates, any material which it held in solution may be deposited as a cement, binding grain to grain. Many substances are sufficiently soluble to furnish cements, but the commonest are carbonate of lime, silica, and certain compounds of iron. It is to these cements that we owe the solid state of many of the rocks of ancient date.

These processes of consolidation disguise the original characters of the rocks so little that it is usually easy to judge the nature of each rock before it was solidified.

There are, however, other changes which affect rocks in a more marked manner, and to these we may briefly allude.

The principal agents which produce these changes are water, heat, and pressure. These agents may act alone or in union with one or both of the other agents. Thus, a rock may be altered by heat, or by heat in the presence of water, or again by heat in the presence of water and under pressure, and as one agent is often rendered more active in union with another, we are likely to find the greater changes which rocks undergo produced where two or more agents work together, and this is most apt to take place at some distance from the earth's surface.

Though two agents often work together, it will be convenient to consider first the mode of operation of each agent, on the supposition that it did its work unaided by either of the others.

Water acts mainly by dissolving materials and depositing them elsewhere. We have considered some of the smaller changes of this nature when discussing the action of weathering upon rocks, and also just now when describing the way in which rocks are cemented. Water produces changes in rocks then by removing material in solution, by adding material to the rock owing to deposit of some substances which the water held in solution, or by altering the composition of the materials already forming the rock. When these changes have occurred in a very marked degree the appearance and nature of the rock may be so altered that its original condition can only be discovered as the result of close study; thus some limestones may be completely changed into rocks composed of silica, and the curious rock known to geologists as serpentine is known to owe its nature to the action of water on a rock quite different to serpentine in its characters.

The effects of heat upon rocks are best studied where igneous rocks have been forced among other rocks in the way described in the last chapter.

Various kinds of changes occur as the result of the heat of the molten rock acting upon the rocks with which it comes in contact. The simplest change is a process which may be described as baking, whereby the rocks in contact are hardened; this change often takes place in the case of mudstones, whereby they become flinty in character.

In the case of sandstones another change often occurs. The outer parts of each grain appear as though melted and then solidified to form a cement between grain and grain. This cement differs from that of an ordinary sandstone, so that the junction between grain and cement is ill-defined.

Limestones are often crystallised as a consequence of contact of igneous rocks, and when pure, a white marble is thus produced, having the appearance of a mass of crystallised white sugar.

In many cases, new minerals are developed as the result of chemical changes in the rocks. Many mudstones in the vicinity of granite are found to contain white rod-shaped crystals. Impure limestones frequently have garnets and other minerals formed in them. The rocks may indeed become completely crystallised, crystals of different minerals being found in the same rock. In such cases it is often difficult if not impossible to find out the exact nature of the rock before the change took place.

There is no doubt that the presence of water plays

an important part in many of these changes which occur
round igneous rocks, but the fact that the changes are
most marked close to the margin of the igneous rock
and gradually become less as one passes away from it,
shews that the heat of the igneous rock is the main cause
of these changes.

The effects of pressure are twofold. In the first
place, certain changes in the physical characters of the
rock are directly due to pressure, and secondly, chemical
changes occur, which may often be due to heat set up as
the result of friction during the pressure.

The physical changes are those which will be noted
here, for the chemical changes are very similar to those
due to the heat given from off an igneous rock and
affecting the rocks in contact with it.

One of the effects of pressure is to squeeze the particles forming a rock, so that they are arranged with
their length at right angles to the direction of pressure,
partly by the particles turning round, and partly owing
to their being squeezed so as to become narrower than
before in the direction in which the pressure is applied
and longer in the direction at right angles to that. The
nature of the change can be illustrated by taking a
number of raisins and pressing them together between
two boards. The raisins tend to arrange themselves so
that the flattened sides lie parallel with the boards, and
as the boards are brought nearer together the raisins
become still thinner, and are squeezed into a longer mass
in a direction parallel to the faces of the boards.

Now, when the particles of a rock have been affected
in this way, all the old planes of weakness, as bedding
planes and joints, may be sealed up, and the rock will
split along a new set of planes, parallel to the flattened

faces of the particles, that is, at right angles to the direction of pressure. A rock of this character is, as stated in chapter III, known as a slate, and the ordinary roofing-slates which are worked so extensively in North Wales are good examples of such rocks. They are usually produced as the result of pressure applied sideways, hence the planes along which slates split are often, though by no means always, inclined at a high angle to the horizontal plane.

When rocks are affected by great pressure they sometimes slip to some extent along planes which lie nearly parallel to each other, and these planes may occur in great numbers in a small thickness of rock. In this way the rock becomes marked by planes of weakness which appear like planes of lamination. Often, after they are formed, they are rucked up into little folds like the laminæ of contorted rocks. These planes have indeed been often mistaken for bedding planes, and the error is a somewhat serious one, for we know that planes of this character may be and often have been produced in igneous as well as in aqueous rocks.

When pressure has caused the rocks to be affected by these and other physical changes, and when heat with or without the assistance of water has allowed the chemical changes which produce recrystallisation of the component mineral particles of the rock to occur, the conditions exist under which those rocks may be produced to which we referred in the third chapter under the name of crystalline schists.

CHAPTER XVII.

CLIMATIC CHANGES.

THE reader who has studied the operations of different geological agents as described in earlier chapters will have grasped the main differences in the action of these agents. It will also have been noticed that the differences in the modes of work of some of these agents are partly due to differences of conditions of climate; thus the wind-erosion of desert regions is distinguishable from the river-erosion of temperate tracts, and this in turn from the ice-erosion of high latitudes.

If then we are able to recognise the particular agent which has been at work in past times from the effects which it has produced, we may get some clue to the nature of the climate in the area under consideration in those times.

Again, we find at the present day that the distribution of animals and plants is largely affected by climate, and this suggests that the study of the distribution of the organisms in the past may also throw light upon climatic conditions.

By combining a study of the physical characters of the rocks and of their included fossils the geologist is in fact enabled to get some idea of the nature of the climate of various parts of the world during the periods of deposition of the various groups of strata.

We shall return to the subject of fossils in connexion with questions of climate in the next chapter. In the meantime let us consider what is meant by the word climate, and also what changes of climate may have occurred in the past.

Climate is at present determined by the amount of heat received by the earth from the sun, which differs in different parts of the world. The more vertical rays of the tropics cause that region to be heated to a greater extent than the regions around the poles, where the rays strike the earth more obliquely. Thus different parts of the earth have different *temperatures*. Temperature is one of the most important factors in the determination of climate, and we speak of a warm climate and a cold climate.

Another important factor is the degree of dryness of an area. We are accustomed to speak of wet climates and dry climates according to the amount of rainfall.

Besides these very important differences in temperature and rainfall, there are many minor causes of variation in climate, some of which are themselves dependent upon variations of temperature and amount of water-vapour supplied to the atmosphere. Among these we may mention the direction of the prevalent ocean-currents and of the prevailing winds, distance from the sea, height above sea-level, and nature of the soil. These in time may affect the temperature and humidity of an area. Climate then is the combined effect of such phenomena.

The geographer of existing times has to take notice of climatic changes, (i) in space, and (ii) in time.

Climatic changes in space. Avoiding minor variations in climatic conditions, we note that at the present day there are marked variations in the climate of the great land-tracts as one travels from the equator towards the poles.

In the tropics the temperature is high and the amount of rainfall great. The tropical climate is hot and wet.

North and south of the tropics are two belts which are characterised by hot, dry climates. In this belt are situate the great deserts of the world. In the northern hemisphere we find the desert of the western United States, the Sahara in Africa, and that of Gobi in Asia, while in the southern hemisphere we note the desert of Atacama in South America, the Kalahari desert of Africa, and the Australian desert. On either side of this belt are the temperate regions in which we in England dwell, marked by fairly high temperatures and moderate rainfall. Still nearer the pole in the northern hemisphere is a region where the temperature is low, marked at first by moss-grown tundras in many places, as in parts of Siberia and a great tract of North America to the south of the Arctic Ocean, while yet nearer the pole is the region of ice and snow determined by low temperatures and by much water-vapour in the atmosphere, which here however falls chiefly as snow, and not as rain.

These great climatic belts running as strips round the earth, approximately parallel to the equator, are marked by notable differences in the work of the geological agents and also in the nature of their assemblages of plants and animals, and it is one of the tasks of the geologist to determine whether similar variations of climate marking different belts of the earth's surface occurred in past times, and if so, how far they coincided with the existing belts.

Climatic changes in time. If we regard any particular area at the present day, we note that the climate of that area varies in course of time. In the first place there are *seasonal* changes. In addition to these seasonal changes, which are readily explained, there are other changes which occur at longer intervals, the causes of which are still obscure. In our own country we are accustomed to spells of successive cold, wet summers extending over a course of years, and alternating with periods of warm, dry summers. Again, we often experience periods of cold, dry winters alternating with periods of warm, damp winters. These longer periods are much more irregular in their duration than are the seasonal periods, though there is a tendency for a period marked by a particular type of climate to recur at roughly equal intervals of time in one man's lifetime. More striking recurrences may come about say once in a century, and it is therefore interesting to the geologist to inquire whether still more remarkable variations of climate may be brought about in any area at long intervals of time—intervals so far apart that the duration of a man's lifetime is insignificant in comparison.

Most people have heard of the Glacial Period, when parts of the British Isles and other regions now enjoying a temperate climate were masked beneath a great pall of ice and snow, and this in itself is a proof that the same area has undergone changes of climate in the course of geological ages. Into the very obscure causes of these changes we cannot here inquire; it is sufficient for us during a preliminary study of the science to know that the changes have occurred.

In later chapters, when we attempt a brief sketch of the geological history of our own islands by piecing together in proper sequence the evidence obtained from a study of the rocks which exist in our islands, we shall see that whereas at some times, as above noted, the climate has been colder than that which now prevails it has at other times been warmer. In some cases we get suggestions of tropical seas, wherein existed reef-building corals, and the shells of mollusks allied to forms now living in tropical seas, at other times the sand-blast of desert tracts sculptured the rocks of the area, and various salts were laid down on the floors of evaporating sheets of water, forming desert lakes. At other times, again, the animals and plants are suggestive of temperate conditions similar to those which now prevail. Again, we find mossy growths due to the existence of mosses allied to those of the sub-arctic tundras, in which are embedded the bones of the reindeer and of other arctic mammals; and lastly, as above stated, the evidences of the existence of great masses of ice and snow indicate that our island has also suffered the severities of a truly arctic climate.

It would appear then that, roughly speaking, indications of all those climates which we noted in a former part of this chapter, as marking different belts from equator to pole at the present day, are found when we study the stratified rocks of our own country.

From these considerations it is apparent that in the past there have been climatic changes in any one area at different geological periods; and we also have evidence of climatic changes in space during the same geological periods, but as these are still very imperfectly known we shall not allude to them at any length in this book.

It was remarked above that climate is at present determined by the amount of heat received from the sun. In a very early period of the earth's history, if we are correct in assuming that the earth has consolidated from a molten state, the temperature of the exterior must have been affected by the heat of the earth itself, which from the point of view of the climate of the present day may be neglected.

Furthermore we are assured that the amount of heat received by the earth from the sun has been diminishing in the course of geological ages.

One would therefore expect to find indications of a gradual diminution of temperature when we pass from the earlier rocks to those which are of later date. So far as the fossiliferous rocks are concerned this does not appear to be the case. The organisms of the earliest known fossiliferous rocks seem to have lived under climatic conditions not markedly different from those which now exist, and the general temperatures of the times when those rocks were formed can hardly have been much higher than those which now prevail. Also, in any area the changes of temperature have not been in the direction of gradual cooling. The rocks of any area shew that at times their temperature was lowered, but that at others it was raised.

This suggests that the period of formation of the earliest rocks containing fossils was one long subsequent to that at which the earth itself came into existence.

CHAPTER XVIII.

FOSSILS.

THE term *fossil* was originally applied to various substances *dug* out of the earth's crust, and not necessarily of organic origin, but though geologists sometimes speak of 'fossil' rain-prints and 'fossil' sun-cracks, the term is usually employed (as it will be in this and the following chapters) in its generally accepted sense. In this sense a fossil is part of some organism or of the track made by an organism, preserved among the rocks of the earth's crust.

An idea often prevails that fossils must be *petrified*, or, as we should prefer to say, *mineralised*. But, though the substance of many fossils has undergone a change since the fossil was embedded in the rock, we find many fossils which have undergone little or no change apart from the destruction of the soft parts of the organism. Many shells have the same composition as when they were embedded, and may even preserve the details of their colour.

Again, an idea of great age is often associated with fossils. But this is misleading. To the geologist a shell which has been buried in the sands last year is as much a fossil as that which has been lying in its rocky grave for long geological ages.

Fossils have been aptly spoken of as "The Medals of Creation." They supply to the geologist the kind of information which the archæologist, who can study human history to a large extent without reference to written records, derives from an examination of coins and other relics of mankind which have been preserved to us from the past.

It will illustrate the importance of fossils to the geologist if we briefly pursue this comparison between fossils and relics of the men of bygone ages.

Human relics like fossils afford us information on two entirely different points. When properly studied we can in the first place use them for the purpose of ascertaining dates. Various objects are now placed beneath the foundation-stones of buildings, and in future, if these objects be discovered, they may be used for finding the period when the building was erected.

But after we have settled the date of their manufacture we may further use the relics to throw some light upon the conditions of the period when they were formed.

We will first take into account the value of human relics for the purpose of ascertaining dates.

Let us imagine that a being could visit our earth who was not aware that man existed. If the being were confronted with a mixture of ancient British, Roman, Saxon, medieval, and modern relics, he would learn but little.

If however he went to work in the same way that the archæologist or geologist has done he would find ere long that cases occurred where ancient British relics existed alone, others where only Roman relics were found, and so on. He might after sufficient experience of this kind come to the conclusion that the ancient British relics belonged to a different period than the Roman ones, though he could hardly be justified in this conclusion until he had long studied the relics, and obtained other information than the mere occurrence of relics of one time unmixed with those of another. For the ancient British types might belong to a different race who lived at the same time as the race who fashioned the Roman types of coins, pottery, and other relics.

Suppose now that our imaginary being discovered in an excavation some such arrangement of relics as is shewn in Fig. 23, where layers of soil are found at different depths beneath the present surface. He would conclude that the upper layers of soil were formed after the lower layers, and that therefore those relics which we have spoken of as Roman were of earlier date than those which we have called Anglo-Saxon; that the Anglo-Saxon in turn were much more ancient than the medieval, and these than the modern.

This is what the geologist has done in the case of fossils preserved in the strata. He is able to find the order of formation of the strata, by knowing that when one stratum or bed was laid on the top of another it must necessarily be newer than that other, and that the fossil remains of the creatures which lived when it was formed and are now embedded in it are also newer than those of the bed on which it was laid down. We have seen that, as the result of much crumpling, a bed once beneath another may be brought above it, but as geologists were careful to work out the order of succession of the beds in regions where they were not very violently disturbed, they were able to find out the actual order in which beds were deposited in different districts, and therefore the relative ages of the fossils included in those beds.

Our imaginary being, looking at the relics which he extracted from the different layers shewn in Fig. 23,

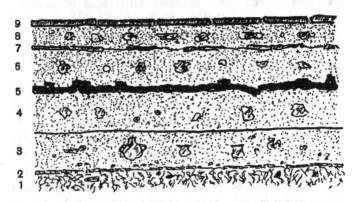

Fig. 23. Section through Old Soils.

9 Modern pavement.   8 Soil with coins of the Georges, modern pottery, &c.   7 Old pavement.   6 Soil with medieval coins and pottery.   5 Site of medieval house.   4 Soil with Anglo-Saxon coins and pottery.   3 Soil with Roman coins and pottery.   2 Roman tessellated pavement.   1 Peat with flint implements.

would soon find that the set of relics in layer 1 differed from those in layer 3, and so with other layers. If he went to another place and examined another excavation, he might find that some of the layers seen in Fig. 23 were here missing, and that other sets of relics not seen in the first excavation were here found. Thus the Anglo-Saxon type of relic might not be represented, the medieval type resting directly on the Roman type; and, on the other hand, a layer might occur between the

medieval and modern layers containing relics of intermediate age.

As our imaginary being continued this work, examining
layers in different districts, finding in one district relics
which were unrepresented in others, but always finding
those relics in the same order, the layer with Roman relics
never above that with those of Anglo-Saxon character,
and similarly with the rest, he would be able to construct
a table shewing the order of appearance of the relics, and
could describe the types characteristic of each different
period, drawing pictures of them, so that others could at
once from the descriptions and pictures distinguish the
Roman sword from the Anglo-Saxon sword, and the coins
of Elizabeth from those of Victoria.

Were our imaginary being gifted with sufficient in-
telligence he would by degrees note that there was a
general advance in the complexity of the relics from the
early stone weapon to the modern rifle, and from the
primitive wooden boat dug from a single tree-trunk to
the recent ironclad. Occasionally there might be a decline,
as from the artistic Roman to the more clumsy Anglo-
Saxon pottery, but on the whole there would be indications
of a general advance in civilisation, so that our being,
coming across relics he had never seen before, might give
a shrewd guess as to the age at which they were fashioned.

So also the geologist, after study of the fossils found
in the beds of the earth's crust, has discovered that on the
whole there is an advance in the complexity of the forms
of life when passing from earlier to later periods, and he
also is able to make a shrewd guess as to the age of a
deposit, even if all the fossils contained therein are new
to him.

Leaving out of account, however, any question of

general advance, the archæologist and the geologist alike can now assign dates to a deposit containing an assemblage of relics, as the result of the knowledge that has been accumulated by those who have worked in the past.

When the archæologist finds pottery of a particular pattern, sometimes marked with the name of the Roman maker, fragments of stones with Roman inscriptions, special types of bronze brooches, necklets and armlets, and coins of the Roman emperors, he knows that he is dealing with a deposit of Roman age, whereas if he discovers sardine-tins, bits of bicycles, and coins of Edward VII, he is aware that he is dealing with a modern accumulation.

So it is with the geologist. In some deposits he will find remains of curious looking animals unlike anything which he is accustomed to see living at the present time, and by referring to books shewing illustrations of the groups of animals that lived in the far past, he is able to discover to what particular period belong the remains which he has found.

Occasionally the archæologist may come across a Roman coin which has accidentally become associated with more modern relics, or again, a coin may be used for some time after the death of a monarch, when another set of coins has been issued. These will, however, be comparatively rare, and if a sufficient number of relics of the actual period when those relics were embedded be discovered the occurrence of stray relics belonging to former periods need not confuse. Once more, this is the case with the fossils, the relics with which the geologist has to deal.

There is yet another point of similarity between the work of the archæologist and the geologist, as regards the light thrown upon dates by the relics with which they deal.

Anyone who found out the order of succession of our
human relics in England would be puzzled if he went to
India, where there would be different types of relics
characteristic of different periods, and he would be compelled to start his research afresh. By degrees, however,
he would be able to get information as to the general
sequence of events in India, and to find out to some
extent what objects were formed at the same time in
India and in England, and this would be rendered still
easier when the order of succession and the types of relics
of those countries lying between England and India were
studied.

Now the geologist also finds that the fossils of beds
of the same date in remote countries are to some extent
different, but he also, as the result of much labour in
studying the fossils of the intervening districts, is able
successfully to compare with some minuteness the rocks
of remote districts, so as to shew what rocks in two
remote countries were formed at the same time.

The first use of fossils then to the geologist is to give
him a clue to the order of succession of beds, and to
compare the ages of the beds of one area with those of
another.

We must now consider the use of fossils as throwing
light on the conditions under which the beds containing
them were formed.

When studying the characters of rocks and their
meaning, we found that it was of the utmost importance
to acquire a knowledge of the changes which are taking
place at the present day, in order to find out the conditions under which the various kinds of rocks were
formed in past ages.

Similarly, our knowledge of fossils would not advance

very far unless we knew something of the manner in which the various groups of existing plants and animals are dependent upon their surroundings.

In the first place we may call attention to some of the differences which mark the plants and animals living on land from those which inhabit the waters of the sea, leaving out of account for the moment the living beings which are occupants of fresh water.

The principal sea-plants belong to simple types of the vegetable kingdom. Anyone who has seen the sea-weeds of the sea-shores will have little difficulty in distinguishing, even without minute descriptions, these plants from the common plants which live upon the land. When we find the impressions of ferns, the roots and leaves of trees, and the seeds of various flowering plants among the strata, we assume that we are dealing with the relics of plants which grew upon the land, for it is unlikely that these plants have changed their mode of existence in modern times, and that having once existed as sea-plants, they have afterwards become occupants of the land.

It must not, however, be assumed that because the relics of land-plants are found in beds, these beds were formed upon the land. Anyone who lives along the coast, especially where rivers enter the sea, will notice that leaves may be blown out to sea, and that seeds of plants, and stumps of trees (or even entire trees), are often drifted out to sea, where they gradually become water-logged and sink. It is clear, therefore, that the remains of land-plants may be found in sea-deposits as well as in land-deposits. If, however, the remains of land-plants can be proved to occur in a deposit in the place where they actually grew, this is proof that that deposit was formed

upon the land, but this is not an easy matter to prove,
for the roots of trees sometimes settle on the sea-floor in
a position imitating that of the roots of the growing tree.

Turning now to a consideration of the land-animals
we note that of those possessing bones, the mammals,
excluding a few, such as whales, which are specially modified for a marine existence, are land-dwellers. Birds may
be embedded equally well in deposits of the land or water.
Some reptiles live in the water and others on the land.
The group which contains the frog and toad also lives
both on land and in the water. The fishes are aquatic,
but may occur in fresh-water or marine deposits.

Of the boneless animals, we may notice first the
shell-forming creatures known as molluscs. The shells
of those which live upon the land are in one piece, like
that of the snail, whereas many of those living in the
waters are formed of two pieces (or valves) like those of
the cockle, mussel, and oyster.

A large number of insects are land-dwellers, as are
also centipedes and other creatures.

The seas contain whole groups of creatures which live
neither upon the land, nor, with unimportant exceptions,
in fresh water. Such are the sponges, corals, sea-urchins,
sea-lilies, lamp-shells, and the group of molluscs which
includes the nautilus.

We noted of land-plants, that their remains were
readily carried into the water-areas, and the same is
the case with land-animals, and therefore the mere
existence of a few relics of land-animals in a deposit does
not prove that that deposit was formed upon the land.
As, however, most marine deposits contain many relics of
animals which existed exclusively in the oceans, it is in
the majority of cases easy to determine the marine nature

of a deposit, even though one or two relics of terrestrial creatures should be found therein.

The relics of marine creatures are more rarely carried inland, though they are occasionally found at some distance from the sea.

These exceptional cases of the transfer of land-creatures into the sea, and of marine creatures on to the land, make geologists very careful to collect all the evidence which they can get, before actually deciding whether a deposit was formed on land or beneath the ocean-waters.

We have had occasion to note that the land is on the whole an area where rock is destroyed, and that the sea is the place where new sediments are chiefly formed. Accordingly, land-deposits are much less common than marine deposits among the rocks of past times, but they are on that account all the more interesting when they occur.

It must be understood that, because of this rarity of land-deposits, our knowledge of the terrestrial organisms of past times is far more incomplete than is our knowledge of the ancient marine organisms.

When we examine a deposit laid down in water to ascertain whether that water was fresh or salt, our task is sometimes a difficult one. The remains of animals having bones are rare, with the exception of fish-remains. Of the boneless animals the molluscs are usually most abundant in sea-deposits as well as those of fresh water. It is true that in most marine deposits some representatives of those groups which exist only in the sea are found, but if they are not found, it does not prove that the deposits are of fresh-water formation. In ascertaining this the geologist must carefully examine the particular kinds of shells, to see whether they are related to those which exist in the sea, or in rivers and lakes. This is a task requiring much special knowledge, and even experienced geologists are sometimes in doubt as to the fresh-water or marine origin of a particular bed, though in most cases the determination is easy.

Passing on to consider more particularly the marine creatures, we need not make a prolonged study of existing forms to find that different creatures live under different surroundings. Some of the shell-fish of our shores are found chiefly in sandy, others in muddy tracts, while others adhere to the rocks. Again, some animals live at greater depths than others, some being confined to the shore belt between high and low water, others being at a depth of a few fathoms, and others again are found hundreds of fathoms below the surface. Our knowledge of the distribution of existing organisms according to depth is still very imperfect, but it is sufficient to give us indications that in past times also waters of varying depth were inhabited by different creatures, and accordingly we are able to tell to a certain extent whether a marine deposit is of shallow water or of deep water origin.

It was noted in the previous chapter that the distribution of organisms is affected by climate, and this is true both of terrestrial and aquatic forms.

The corals which form coral-reefs are at present confined to warm seas, and many of the molluscs of the tropics are different from those which inhabit the seas of higher latitudes. Accordingly, by comparing the fossils of any formation with their recent allies, we may get some insight into the climatic conditions under which that formation was deposited, though as this is a task fraught with much difficulty we cannot pursue the subject here.

It has been remarked before that when examining the strata formed in past ages, the included fossils become less like the hard parts of existing organisms as we work backwards from the more recent to the more ancient deposits. Accordingly, inferences as to the conditions under which deposits were formed are much more readily made in the case of the more recent deposits where the organisms are closely allied to existing forms, than in the more ancient sediments where the differences from living forms are very marked.

In conclusion, as the evidence regarding conditions of formation of beds derived from study of the fossils is so often obscure, it is necessary, in order to discover those conditions, to study the characters of the rocks themselves as well as those of their fossils, checking one kind of evidence by the other.

CHAPTER XIX.

THE PRINCIPLES OF CLASSIFICATION OF ROCKS ACCORDING TO AGE.

WE have noticed that there are two great tests by which we are enabled to discover the relative ages of the strata of any area.

The first of these is by observing the order of succession of the beds, for unless they have been so disturbed that their original order has been inverted, the newer beds repose on the older. (One bed is said to rest on another not only when the two are flat, but when they are tilted at any angle between the horizontal and the vertical.)

The second test is that of the included fossils. From what we have said in the last chapter it is clear that as the result of our information gradually acquired by observation, we are able not only to say that certain beds in one area are newer than others on account of the nature of the fossils in each, but we are also enabled to compare beds of the same age in two different areas by noting similarities in their included fossils. How far the beds of remote areas can be stated to have been formed at the same time is a matter which is still under discussion. In this book we do not propose to touch on the geology of other countries, nor shall we enter into a minute descrip-

tion of the smaller divisions of the strata of our own
islands; accordingly we need not go further than to state
that there is now general agreement as to what beds of
the various parts of our own country were formed at the
same great geological periods.

In writing an account of any events with reference to
the times at which they occurred, we must have some way
of dividing our time. For ordinary purposes we divide
time into seconds, minutes, hours, days, weeks, months,
and years. For other purposes we may have less regular
divisions. · Thus, in writing a history of England, it is
convenient to separate that history into chapters, where
each chapter is devoted to the records of a series of events
which are distinguishable from the events recorded in
another chapter by some important change, such as the
end of a monarch's reign. These chapters then will not
refer to periods of equal length, and as pointed out in the
Introductory Chapter, the chapters dealing with the
history of one country would not cover the same periods
of time as those covered by the chapters dealing with
another country's history.

Now in the geology of every country we find indications of a change of unusual amount having ever and anon
occurred, and these periods of change are convenient as
enabling us to divide the geological history of that country
into a series of chapters each of which is separated from
the next owing to the occurrence of one of these marked
changes.

If after a long period during which an area like our
own has been occupied by sea, on the floor of which
deposits were laid down, the area is uplifted and formed
into land, the changes which have been described as
occurring on land-tracts will now take place. These

changes are mainly of the nature of destruction of rock by erosion; when the tract is once more covered by sea, a new set of deposits will be laid down on the tilted and eroded surface of the older ones, giving rise to an unconformity, as described in chapter XI. An unconformity then may be and often is utilised in any area for separating two chapters of our geological history. It must be distinctly recognised that this unconformity can only be utilised in limited areas for this purpose. When one tract is land another must be occupied by sea, so that during the period of erosion in one area sediments are being formed in adjoining regions, and these *breaks* in the rocks are therefore of limited extent. At the present time erosion of the already formed rocks is taking place on our continents, so that in the future, apart from a few scattered terrestrial deposits which may owing to exceptional causes be preserved, the present continents will yield no strata representing the present period, whereas sediments which do represent this period are now being laid down on the existing ocean-floors.

Often we find that a group of strata separated by unconformities from the groups above and below are also marked by certain recognisable features as regards their fossils by which they are distinguishable from the overlying and underlying groups.

But we need not separate our larger chapters only when unconformities are found.

If we get a group of rocks having some marked characters in common, whereby they are to be distinguished from the groups above and below, we may also treat these rocks in a distinct chapter.

The groups of rocks which are separable in the above ways, so that we have spoken of them as useful for dividing our geological volumes into chapters, are spoken of as Systems of strata.

There are, however, larger divisions of the groups of strata than these, which though of doubtful value are generally used. We speak of the Primary, Secondary, and Tertiary divisions of rocks. These may be compared to the three great volumes of our earth's history.

In the Primary Division we place about half-a-dozen systems, in the Secondary Division are only three, and in the Tertiary Division are a number which vary greatly according to the classification we adopt.

Just as the chapters of our human histories comprise periods of very unequal duration, so it is with the chapters of our earth's history. The rocks of one system are often much thicker than those of another, and may have taken a very much longer time for their formation.

The time taken for the formation of the rocks of any one system is unknown to us, and we can get but the vaguest notion of the durations of geological time. We speak of the earth as being millions of years old, but it is very difficult to grasp the idea of a million years. Perhaps the best idea of geological time which we can get is by comparing it with historic time.

Written history goes back for a few centuries, but it nevertheless covers a period many times longer than that of a single lifetime. The first appearance of man in the strata occurred long before the period of our first written records; we may perhaps very roughly compare the difference between the human period as a whole, and that of the period of written history, with the difference between the latter and the duration of a single life. But the appearance of man on the earth was, geologically speaking, an affair of yesterday as compared with the

times which have elapsed since the formation of the earliest
known fossiliferous strata; and, as has been pointed out
in the last chapter, this is perhaps but a fraction of the
time which has elapsed since the earth came into being.
In this way one gets a slight idea of the vast periods of
time which are covered by the history of the earth as
a whole.

Let us now return to the geological Systems. Each
of these is subdivided into two or more Series, and the
Series are again subdivided into Stages. The rocks of
one series have something in common whereby they are
distinguished from those of another, and the same is the
case in a minor degree with the rocks of a stage.

We will now write out a table of the strata of the
British Isles, which we shall follow in the descriptions of
those rocks given in the succeeding chapters.

		Systems
Tertiary	{	Recent
		Pleistocene
		Pliocene
		Miocene
		Eocene
Secondary	{	Cretaceous
		Jurassic
		New Red Sandstone
Primary	{	Carboniferous
		Devonian
		Silurian
		Ordovician
		Cambrian
		Precambrian

It must not be supposed that there is a general
agreement among writers as regards the details of this
classification. It is to be specially noticed that the later

divisions, such as Recent and Pleistocene, assume importance on account of their modern formation. Were they very ancient, the Tertiary rocks of Britain would probably be separated into two systems at most.

Before discussing in order the characters of the rocks of the principal divisions in the British Isles, and outlining the story as to the conditions of our area during different periods which these characters tell us, we must give a short account of the geographical distribution of the strata on our islands.

The older rocks of Great Britain are found mainly in the north and west, and the newer ones in the south and east. Accordingly, the dip of the rocks is on the whole from north-west to south-east.

The structure of the island is far from being so simple as might be supposed from the above statement. Numerous minor folds cause isolated patches of older rocks to appear among the newer ones, or of newer among the older strata. Faults and unconformities further add to the complex structure of the island.

It is however convenient for us at present to ignore the numerous variations from the simple arrangement noted above, and to consider the beds as though they were arranged in a regular manner, so that those of the north and west dip under those to the south and east.

As the result of this general structure the oldest rocks of Great Britain are found most extensively in the Highlands of Scotland, but patches are also found in England and Wales, one of the most interesting of these being in the island of Anglesea.

In Fig. 24 we have given a simplified section of the strata between Anglesea and London, where the beds do succeed one another in their proper order, when viewed on a large scale. It is true that in Anglesea many newer beds are let down among the oldest rocks, but this is a very local variation.

In that part of the section which lies between the Menai Straits and the border-line between Wales and England, the hilly country of North Wales is occupied by Lower Primary rocks. On the border-land tract the Upper Primary rocks are seen. The country between Worcester-shire and Buckinghamshire is occupied by the Secondary rocks,—the older to the north-west, and the newer to the south-east, and in Middlesex we find the Tertiary rocks developed.

If we look at a geological map of England and Wales we shall find that the Lower Primary rocks occur chiefly over the greater part of North and Central Wales and over much of the west of South Wales, and also in the Lake District of Cumberland and Westmorland.

The Upper Primary beds form the chief part of the Pennine Hills from the extreme north of England to Derbyshire, and also tracts of the lower lying ground

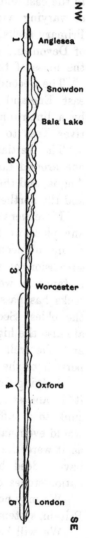

Fig. 24.
Geological Section from Anglesea to London.

1 Precambrian Rocks.	2 Lower Primary Rocks.
3 Upper Primary Rocks.	4 Secondary Rocks.
5 Tertiary Rocks.	

to the east and west of these hills; they also form a fringe of varying width to the east and south of the Lower Primary rocks of Wales, and occupy a considerable part of Devon and Cornwall, while an interesting patch forms the mass of the Mendip Hills.

The Secondary rocks occupy the greater part of South-east England, being separated from the Primary beds by a line running on the whole from the mouth of the river Tees to Torbay.

The Tertiary rocks form two important patches, the one around London, with some isolated portions in East Anglia, and the other in parts of Hampshire and Sussex, and the northern half of the Isle of Wight.

If we were to sink a shaft through the newer rocks of any place we should find older beds underneath, though, owing to unconformities and for other reasons, the whole succession of strata would probably not be found. For instance, a well in London sunk through the Tertiary rocks has passed through those of Secondary age. But the oldest Secondary rocks are there absent, as indeed are also the highest members of the Upper Primary rocks, and the well ended in strata belonging to the lower portion of the Upper Primary rocks.

Judging from what we know of the geology of Europe, it is almost certain that in every case, if a shaft were sunk to a sufficient depth in any part of our islands it would eventually reach the Precambrian rocks, which are as it were the foundation-stones of our island; and they have indeed been spoken of with reason as the 'Foundation-stones of the Earth's Crust.'

We may now proceed briefly to describe the strata of Britain, system by system.

We will begin with the older systems. In so doing,

we have, it is true, to consider the more obscure rocks before those more modern beds which contain fossils most nearly resembling the existing plants and animals, but it is better, notwithstanding, to learn our earth's history by taking the events in the order in which they occurred.

In the case of each group we will give a general description of the rocks which compose it, followed by a brief account of the main characters of the fossils. This will enable us to discuss in general terms the conditions under which the rocks of the group were formed in our country. Lastly, a short allusion to the effect which the rocks have had upon the characters of the surface features of the country will not be amiss, as enabling the reader to recollect the general characters of the rock, and shewing him how important rock-structure is as a factor in determining various types of scenery.

CHAPTER XX.

PRIMARY ROCKS.

1. THE PRECAMBRIAN AND LOWER PRIMARY SLATE ROCKS OF BRITAIN.

A. *The Precambrian Rocks.* Though, as already stated, these rocks occur in Anglesea and in a few other places in England and Wales, their chief development in Great Britain occurs in the Highlands of Scotland. Here they form two very distinct groups, as shewn in Fig. 25, which is a rough section through part of Ross-shire from north-west to south-east.

The section shews two very distinct types of Precambrian rocks,—a lower group consisting of *gneisses and schists,* and an upper one formed largely of *red sandstones.* The upper group rests unconformably upon the lower one, and the Cambrian beds rest also unconformably upon lower and upper Precambrian rocks.

The sandstones are stratified rocks, but the gneisses and schists are in their present condition very unlike ordinary strata. From what has been said in chapter xv. the reader will gather that these gneisses and schists have undergone alteration since the rocks were formed. We further learn from examining pebbles of the gneisses and schists which are found in coarse pebble-beds at the base of the upper group of the Precambrian rocks, that this alteration took place before the formation of the rocks of the upper group, for the pebbles are fragments of rocks which possess the same characters as the main masses of gneisses and schists now exhibit.

This type of highly altered rock is one which is found very widely distributed in most parts of the world where Precambrian rocks occur. Further, though the original nature of these altered rocks is in many cases impossible to discover, we have proofs that large masses of the rocks are of igneous origin, and that they were consolidated beneath the earth's surface, and afterwards exposed by erosion. The abundance of rocks of this type suggests that igneous action and alteration of the type which produces gneiss and schist were exceptionally rife in Precambrian times, but the question is still so obscure that it is better not to enlarge upon it here. We will merely remark that the conditions which then prevailed may have differed to such an extent from the existing conditions of the earth as to have produced differences in the general characters of the rocks.

Fig. 25.

Section across part of Ross-shire.

3 Cambrian sandstones and limestones.   2 Precambrian red sandstones.   1 Precambrian gneisses and schists.

The nature of the upper Precambrian rocks of the Highlands is less obscure than that of the gneissose and schistose rocks. They are clearly stratified, but in the case of the particular strata found in the Highlands the evidence suggests that they may be terrestrial beds. Other Precambrian stratified rocks occur in other parts of the world. There is another difficulty connected with these rocks. Fossils swarm in many of the Cambrian strata, but some geologists maintain that no undoubted fossils have been found in the Precambrian rocks. This may simply be due to our ignorance of these very obscure rocks, and most geologists look forward in expectation of the future discovery of undoubted fossils in them.

From what has been said, it will be gathered that the history of these foundation-stones of our island has still to be written. At present much connected with them is shrouded in obscurity, and this being so, it is best to confess our ignorance of the exact mode of formation of the bulk of these rocks, and therefore of the evidence which they give as regards the conditions which prevailed during their formation. Let us pass on to a description of the less obscure rocks of Lower Primary age which are found above the foundation-stones.

B. *The Lower Primary Slate Rocks.* The general distribution of these rocks in Britain has already been indicated. A study of this development in the upland regions of Wales will give us a good idea of their main characters.

Some writers have divided them into two systems, others into three. The lower portion is usually spoken of as Cambrian, and the upper as Silurian, while those who recognise a middle division term it Ordovician.

We will not take account of these divisions in detail,

but simply speak of the rocks as a whole as the Lower
Primary Slate Rocks. This title gives a general notion
of one of their characteristics as developed in our country.
They are largely *slates*, and indeed the principal roofing-
slates of the world are extracted from the Lower Primary
Rocks of Wales.

We have already seen that the production of slaty
structure is due to change taking place in rocks subsequently to their formation, so that, although the great
group of rocks is marked in our country by possessing
the slaty structure, this was not produced at the time
of their deposition.

A study of the rocks shews that in this country they
consist chiefly of mudstones and sandstones with comparatively few limestones. One of the great features as
regards the rocks which form the group is the abundance
of material which on examination proves to be of volcanic
origin. Both lavas and volcanic ashes are largely represented among these rocks. In some cases, as in Cumberland and Westmorland, we find many thousands of
feet of lavas and ashes unmixed with ordinary sediments,
and similar volcanic rocks are abundant in many parts
of Wales, as in Carnarvonshire and Merionethshire, while
it is clear that the finer volcanic dust has in many cases
contributed largely to the formation of the mudstones.

The Lower Primary Fossils. One of the most remarkable points in connexion with the fossils of these
ancient rocks is their great variety. The animals possessing
bones do not appear until towards the end of the period,
and these are the most lowly of the bony creatures—the
fishes; but all the main groups of the boneless creatures
are represented, the greater number of them actually in
the earliest of the Lower Primary Rocks. The contrast

between the doubtfully fossiliferous Precambrian rocks and the richly fossiliferous rocks which succeed them is indeed curious. We cannot suppose that the rich and varied assemblage of creatures which is thus abruptly met with came into sudden being. All the knowledge which has been acquired by students of life during the last half century goes to shew that this complex assemblage can only be accounted for on the supposition that long ages elapsed before the formation of the rocks containing the fossils of these many kinds of creatures; ages during which there was a gradual increase in the variety and

Fig. 26 A. Graptolites.

complexity of the animals. We have previously noticed other reasons for suspecting this lapse of ages before the formation of these early fossiliferous rocks, and we now see that the nature of the fossils in these rocks gives additional support to our suspicion.

Let us now notice to some extent the characters of the fossils of the slaty Lower Primary rocks. It is not of much use writing a long account of these, apart from details as to the difference of structure between many of the ancient groups of animals, and their nearest living relations; this could not be done without much study of animal structure. The mere general appearance of a fossil is not of much value to the proper study of our science. Indeed, vague descriptions of fossils are worse than useless, for they tend to encourage a fault which is

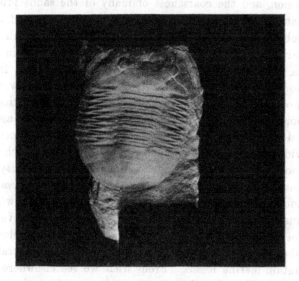

Fig. 26 b.   A Trilobite.

very baneful to the proper progress of science, namely, a
want of exactness. We will merely remark therefore that
whereas some of the fossils found in these early rocks
are not dissimilar to the hard parts of existing organisms,
others are very different from anything now living. In
Fig. 26 are representations of two groups which are very
characteristic of these early slate rocks. The *graptolite*
(A) is a distant relative of the sea-pens which are often
found on our shores, while the *trilobite* (B) is remotely
related to the living crabs and lobsters.

*Conditions under which the rocks of the Lower Primary
slate group were formed.* The abundance of sands and
muds shews that a tract of land was not very far from our
area when these slaty rocks were deposited: for the
sand and much of the mud was derived from land by
erosion, and the coarseness of many of the sands proves
that the coast was near during parts of the period, and
probably at no time very distant. The fossils which are
marine also tell us that the deposits were formed in the
sea, and not in lakes. Further, we should judge from
the fossils that the sea was in many cases shallow, and
in others of greater depth, though here our information is
somewhat vague. That this tract of ocean was again
and again during this period occupied by volcanoes is
evident from the abundance of volcanic rocks which are
associated with the ordinary sediments. The rarity of
limestones is no doubt partly due to the nearness of
land, but also to the quantity of volcanic dust which
must have been showered far and wide from these vents.
For instance, the rocks on the top of Snowdon are volcanic
ashes which were piled up on the sea-floor, and therefore
contain marine fossils. From what we see elsewhere we
should judge that, but for the presence of a volcano in

the vicinity, these rocks on the extreme top of Snowdon would have been made up of limestone and not of volcanic fragments.

We know little of the climate which prevailed in this area during the period. As so many of the fossils are only remotely related to living forms, they throw little light on the subject, and the composition of the rocks affords little assistance. Some of the limestones which are found, though rarely, are largely composed of corals which built reefs in these early days. At the present day reef-building corals are confined to warm seas, and so far as we dare judge from analogy, we therefore hazard the suggestion that during these parts of the period the climate was warm. This is to some extent borne out by the shells which are found in these coral-bearing limestones; some of these recall the shells now living in the tropical or sub-tropical seas.

The sea-floor must have sunk on the whole during these times. We had reason from the occurrence of an unconformity at the top of the Precambrian rocks to judge that land rested over our area at the end of Precambrian times. As this land sank the first shallow water deposits were laid down upon it. But the slaty group of rocks has a thickness of many thousands of feet; in some places— apart from volcanic matter—there must be at least 30,000 feet of sediment. This clearly could not have been formed unless the old land sank to that extent, and while the sea-floor was sinking the piling up of sediment went on at much the same rate, thus ever keeping the sea-floor fairly shallow, though sometimes of greater depth than at others. At the end of these times any inequalities which had existed would have been largely buried beneath the blanket of sediment which was piled up, and the shallow sea-floor at the close of the period probably possessed a fairly level upper surface due to the sediments.

Scenery of the districts occupied by these old rocks. In considering the effect of these ancient rocks upon the present surface features we may regard both the Precambrian and the Cambrian rocks as forming one group. The highlands of our island are largely occupied by these rocks, which have indeed to a large extent determined their highland nature. Most of the rocks are hard, and there is also a considerable variety of hardness, which has given rise to much diversity of feature. The crystalline Precambrian rocks are hard largely on account of the changes which they have undergone. They stand out forming the prominent mountains of the Scotch Highlands. The slaty Lower Primary rocks are hard, partly owing to the changes which impressed their slaty structure upon them and partly because they contain so much volcanic and other igneous material. Snowdon and its attendant peaks in North Wales are carved from these volcanic rocks, as are also Scawfell and the adjoining summits in Cumberland, while Plynlimmon in Wales and Skiddaw in Cumberland are composed of sediments which have been hardened by the pressure which they underwent after their deposition.

The rocks were also greatly folded, and are more often tilted than level, and accordingly the terraced features which we shall subsequently note as being common in rocks which are level or nearly level are here rare.

Again, the characters of the rocks (assisted to some extent by the present climate of the uplands) forbid the accumulation of a rich soil, and great tracts of these upland regions are occupied by coarse grass and heather,

while in others naked rocks project on the summits of
ridges and the sides of the valleys.

One other point must be mentioned. The hilly
character of these tracts has, at no long distant date,
allowed the accumulation of snow to such an extent that
glaciers crept down the valleys, and in some cases great
masses of moving ice covered hill and dale alike ; this
growth of land-ice produced its effects upon the scenery,
but those effects will be considered when we, in a later
chapter, discuss the conditions of our country during that
age of ice.

CHAPTER XXI.

THE PRIMARY ROCKS.

2. THE UPPER PRIMARY SYSTEMS.

A. *The Devonian System.* Towards the end of the last period important movements commenced in our area, as the result of which physical conditions very different from those which existed during the formation of the group of slaty rocks prevailed in the British area during the earlier portion of Upper Primary times, when the rocks belonging to what is termed the Devonian System were laid down.

As the result of these movements, which affected the southern part of our tract in a less degree than the northern, two distinct types of deposit were laid down; one in the south may be spoken of as the Devonshire type, while the other, which is represented by rocks in Wales, the Welsh border-land, and Scotland is known as the Old Red Sandstone type.

The rocks of the Devonshire type are not unlike those of the slaty groups of Lower Primary times in general characters, and although their fossils differ in some points from those of the earlier period, there is here also a general resemblance. One of the most striking differences is the fair abundance of limestones, formed partly of reef-building corals, in the Devonian rocks of Devonshire. These

Devonian rocks of marine origin are found in North Devon, and also in South Devon and parts of Cornwall.

The Old Red Sandstone type is more interesting on account of its somewhat exceptional characters, and we must treat it in rather more detail.

It is found in three great patches and some smaller exposures. The most southern of the large tracts is in south-east Wales and parts of the border counties. The next occupies part of the lowlands of Scotland in the basins of the Forth and Clyde, and the third occurs in the north-east of Scotland and the Orkney Islands.

In all of these tracts the main mass of rock consists of beds of red sandstone, often containing extensive pebble-beds, but in the Scotch area a very considerable development of volcanic lavas and ashes occurs in the heart of the Old Red Sandstone.

In North Wales and Cumberland rocks of this age are absent, and the rocks of the next system (the Carboniferous) rest unconformably upon the upturned and eroded edges of the Lower Primary strata. This was one of the effects of the movement to which allusion has been made. Tracts of land with a general east-north-east and west-south-west trend, as proved by the direction of the axes of the folds of the Lower Primary rocks, were brought above sea-level and underwent erosion in Devonian times, and these exercised an effect upon the nature of the Old Red Sandstone rocks, as will be noticed more particularly below. The movement also had an effect upon the Lower Primary strata (and also to some extent upon Precambrian rocks). The slaty structure which the earlier rocks now possess was partly impressed upon them as the result of the side pressure exerted on those rocks during the movements.

Fossils of the Old Red Sandstone. The remains of boneless animals are not abundant in these sandstones, and the most interesting fossils are the relics of fish which are so well-known owing to the luminous writings of Hugh Miller. These fish were in many cases encased in a kind of bony armour outside the body. Some of them were of considerable size. The nearest existing allies are found in certain rivers, and it has been suggested, from the abundance of these fish and the rarity of boneless animals in the Old Red Sandstone, that the deposits were laid down in fresh water, but we will return to this question immediately.

Conditions of formation of the Old Red Sandstone. The movements which we noted above caused the formation of ridges which underwent erosion as the land was uplifted. The intervening hollows at the same time became depressed, and into these hollows, which were occupied by water, the materials which now form the sandstones were carried from the adjacent land. The coarseness of grain of the rocks and the frequency of pebble-beds suggest that the erosion of the ridges was caused by rapidly flowing streams such as occur in hilly regions. Could we have gazed at the scene, we should probably have seen long, hilly peninsulas, and perhaps at times islands, trending east-north-east and west-south-west, and separating these shallow gulfs with chains of volcanoes in places, in which the Old Red Sandstones were laid down. The question as to whether these water areas were of the nature of lakes or arms of the sea is still open, though many geologists maintain, from the striking contrast between the Devonshire and Old Red Sandstone type of Devonian strata, and for other reasons, that the Old Red Sandstone was laid down in fresh-water areas.

Sir Archibald Geikie further suggests that the areas in which different tracts of Old Red Sandstone occur formed different sheets of water, and he refers to the Welsh Lake, Lake Caledonia, and Lake Orcadie, in which he supposes that the deposits of Wales and the border-land, the central lowlands of Scotland, and the north-east of Scotland were respectively laid down.

As the Old Red Sandstone has a thickness of about 20,000 feet, great masses of the rocks of the bounding lands must have been eroded. In the central lowlands of Scotland about 6000 feet of volcanic rocks of different types are associated with the sandstones.

Scenery of the districts occupied by Devonian rocks. The rocks of Devon and Cornwall give rise to features somewhat resembling on a smaller scale the hilly tracts of Wales. The Old Red Sandstone ground is usually hilly, the surface of the country being often rounded and marked in places by rich, warm soils. The volcanic rocks of the central lowlands of Scotland, like those of the Lower Primary groups of Wales, have resisted erosion more successfully than the softer sediments which surround them, and they now stand out as hill ranges, the Ochil, Sidlaw, and Pentland Hills.

B. *The Carboniferous System.* As the result of erosion of the land-tracts, and silting up of the water areas by sediment during Devonian times, a comparatively level surface was produced over the British area at the beginning of the period which witnessed the laying down of the strata of the succeeding Carboniferous system. It is true that some parts were submerged during a downward movement of the surface, and accordingly the lowest strata of the Carboniferous system in some parts of our area were formed before those of other portions. But leaving out of account these local details, we may separate the Carboniferous rocks into lower and upper divisions. The rocks of the lower division are marked by the abundance of limestone in many parts of the area, with frequent beds of shale, which are specially common towards the base and summit of this division, while the rocks of the upper division consist especially of sandstones with some shales, and in the later portions they contain those important coal-seams to the existence of which our country owes so much of its prosperity.

Beginning with a description of the rocks of the lower division, we may note that the limestones are specially thick and pure towards the centre of England, being very well developed in the Mendip Hills, and particularly in Derbyshire. These limestone beds constitute what is often spoken of as the Mountain Limestone, on account of the way in which the country which is composed of it frequently forms hilly ground. The limestones are usually fine grained and of a grey-blue colour. In Derbyshire they are at least 3000 feet in thickness and are succeeded by many hundreds of feet of impure limestones, shales, and sandstones, also belonging to the lower division of the Carboniferous system. The limestones of this county contain some lavas and ashes which were formed by submarine volcanoes.

The limestones are almost entirely formed of the remains of organisms of various kinds. Some beds are largely made of shells, others of reef-building corals, but large masses of the limestone are formed of portions of creatures known popularly as sea-lilies, though they are not plants but animals related to the star-fishes of modern seas, but usually attached to the sea-floor by a column formed of a number of bead-like joints, each having a hole in the centre. These limestones are spoken of as *encrinital* or *crinoidal* limestones, as the sea-lilies which form these are also known as crinoids, and their fossil relics as encrinites.

The Upper Carboniferous Rocks do not differ very greatly in different parts of the country. The lower beds consist of very coarse sandstones, whose greatest thickness, on the borders of Lancashire and Yorkshire, is 3500 feet. These sandstones are called Millstone Grit, as some of the beds have been used for making millstones. Above the Millstone Grit comes the most important division of the Carboniferous rocks, known as the Coal Measures. The sandstones of the Coal Measures are usually finer than those of the Millstone Grit. Shales also occur. The most important beds, however, are the coal-seams, which, though never very thick, often extend for long distances without much change in thickness and in their characters. These coal-seams often rest upon masses of rock known as *underclay*. The underclays are unstratified muds with many roots and rootlets running through their mass.

The coal measures where thickest, namely in South Wales, measure about 11,000 feet from top to bottom, though the united thickness of all the coal-seams is only a few score feet.

These coal measures, owing to movements which occurred after their formation, are now found in basins scattered over the country. The principal basins are, firstly, that of the great central coal-field of Scotland; secondly, a group occurring on either side of the Pennine chain which is itself formed of Lower Carboniferous rocks; the group consists of the Northumberland and Durham coal-field and that of Yorkshire, Nottinghamshire and

Derbyshire, to the east of the Pennines and of the Cumberland coal-field and that of South Lancashire and North Staffordshire to the west of that chain; thirdly, several scattered fields in central England, among which are those of South Staffordshire, N. Wales, Leicestershire, and Worcestershire; fourthly, the important coal-field of S. Wales, that of the Forest of Dean, and the Bristol and Somersetshire coal-field; fifthly, a few small coal-fields in Ireland.

The Fossils of the Carboniferous Rocks. The British fossils of Carboniferous times shew that the Lower Carboniferous beds were formed under surroundings which differed in many respects from those which marked the formation of the Upper Carboniferous strata. The prevailing Lower Carboniferous fossils are the relics of ordinary marine creatures, both backboned and without backbones. Those of the Upper Carboniferous beds which are most noteworthy are plant remains.

The animal fossils in many respects recall those of earlier rocks, with differences in detail. The great group of trilobites which was abundant in Lower Primary times and fairly common in Devonian times is poorly represented in the Carboniferous rocks, and there disappears. Corals, sea-lilies, lamp-shells, and molluscs are abundant. Of creatures having bones fish are fairly common, and the group of amphibious animals to which belong the frog, toad, and newt appears in the rocks of the Carboniferous system.

The plants require further consideration. They do occur in Lower Carboniferous rocks in Britain, but are most abundant in the upper beds, and especially in the coal-measures.

Plants are found even in the Lower Primary rocks,

and are not uncommon in the Devonian rocks, but the Carboniferous period was marked by the abundance and luxuriance of the flowerless plants, of which the groups represented by the modern ferns, club-mosses and horse-tails play the most important part, not only on account of their abundance, but also because of their extraordinary size. We have tree-ferns now growing in warmer parts of the world, but the modern horsetails and club-mosses are usually of lowly growth, whereas their Carboniferous relatives, like the ferns of that period, often assumed the height and girth of forest trees. A strange sight must have been presented by these forest growths, very different from the appearance of a modern forest. Perhaps nothing would more vividly shew the changes which had taken place in the course of geological ages than a tour through these ancient jungles, if we could make it, fresh from a visit to a modern forest. The title, "Age of flowerless plants," applied to the Carboniferous period, certainly calls attention to one of the most interesting points in con-nexion with that period.

Conditions of the Carboniferous Beds. The Lower Carboniferous rocks of Britain were mainly formed in the open sea, though in the south of Scotland we find some freshwater deposits of this age. That the sea was tolerably clear through much of Lower Carboniferous times is proved by the abundance of fairly pure limestones, but it was probably of no great depth, as indicated by the frequent occurrence of limestone formed of reef-building corals, and by the mixture of muddy and sandy sediment which occurs with the limestones as we pass from central England and Ireland in a northerly direction.

The climate, as far as we can tell from the fossils, was probably somewhat warm.

The great change in the characters of the Upper Carboniferous beds marks a difference in the conditions under which they were deposited from those which existed in Lower Carboniferous times. The replacement of limestones and shales by massive sandstones marks the nearer approach of the shore-lines to the British area. The Millstone Grit may be regarded as formed of great sandbanks heaped up off the coast-line in very shallow water; and at the end of Millstone Grit times, our area by the filling up of the Carboniferous water-tract with sediment appears to have been of the nature of a flat plain of sediment at or near sea-level. This plain must have sunk during Coal-measure times, to allow the thousands of feet of Coal-measure sediments to be laid upon the Millstone Grit, and the sediment was poured into the water area as fast as its floor sank, so that the Coal-measure sandstones were always laid down in very shallow water, again and again raising the surface to sea-level. When this happened, and a pause from the downward movement occurred, the plants of the period gained a footing on the flats, and the forest growth extended over the great tracts of the plain. The upper deposits of that plain rotted and gave rise to a soil, in which the roots of the flowerless trees grew, while year by year the dead vegetation was piled over the top of the soil, giving rise to a matted mass of vegetable matter—at first, no doubt, of a peaty consistency—to be turned in the course of after ages into coal. Thus was produced a coal-seam resting on its old soil of underclay. When the period of pause was succeeded by one of downward movement, the seam of vegetable matter would be again sunk beneath the waters, to be once more buried by sandbanks, and when another pause occurred, the plants would once more spread over the surface to give rise to another forest, and ultimately to another coal-seam, at a higher level. Thus, as the result of slow sinking of the floor of a shallow water area, interrupted by occasional pauses lasting for a long time, the Coal-measures were gradually piled up.

The climatic conditions of the Upper Carboniferous times have been the subject of much discussion. The exact temperature cannot be determined; it was probably fairly high, but not extremely so. Geologists, however, seem to be agreed that the climate was *humid*. The abundant growth of luxuriant plant-life of the type which we have noticed as prevalent in Carboniferous times demands an abundant supply of moisture, and we must therefore suppose that the amount of rainfall was fairly high.

It will be noticed that the coincidence of three important sets of conditions was required to form extensive coal-fields. In the first place, as the result of gradual downward movement, accompanied by deposits of sediments, and *with little movement of the kind which produces folding of the beds*, a flat tract at or near sea-level was produced over an exceptionally large area, not only in Britain, but also in many other parts of the world. Secondly, the climatic conditions seem to have been favourable for the growth of a particular type of plant over wide areas, and thirdly, that type of plant was the prevalent one then existing. Such conditions never seem to have occurred on so large a scale either before or after Upper Carboniferous times, and accordingly, though coal-seams are found among the beds of other geological systems, they never assume such importance in other times as they did in those of the Upper Carboniferous period, as regards number of seams, thickness of individual seams, and area over which the accumulations of coal extend.

Scenery of the Carboniferous Rocks. As these rocks in Britain are frequently found but slightly inclined and often horizontal, and as easily eroded deposits like shales often occur between more resistant rocks as limestones and sandstones, these latter usually stand out in long lines of cliffs or scars, such as occur abundantly in Derbyshire, West Yorkshire, and elsewhere. The limestone scars of Lower Carboniferous times usually appear white from a distance, for little soil grows on them, and the naked rock is usually visible. They are frequently riddled by swallow-holes and subterranean caverns. The grit scars are darker in colour, and often alternate with tracts of boggy or heathery moorland, where a thin soil has been formed, as may be seen in many parts of the Peak district of Derbyshire, and on the borders of Lancashire and Yorkshire.

The Coal-measures, formed on the whole of more easily eroded sandstones than those of the Millstone Grit, are often marked by lower ground usually somewhat barren, and occupied by coarse grass and boggy growth.

CHAPTER XXII.

THE SECONDARY ROCKS.

THE NEW RED SANDSTONE PERIOD.

THE period for which we have adopted this name, which has been given on account of the nature and colour of its most abundant deposits, contains the rocks of two different systems, known to geologists as the Permian and the Triassic systems. The earlier rocks, belonging to the Permian system, are really included among the Primary rocks, while those of the later Triassic system form the lowest members of the Secondary rocks.

In Britain, however, the rocks of these two systems present so many features in common, and the Permian rocks, on the other hand, are so sharply separated from those of Carboniferous age, that it will be convenient to regard the rocks of Permian and Triassic systems alike as referable to the Secondary rocks of this country.

Great were the changes which occurred in this area at the end of Carboniferous times, and striking therefore is the contrast between the rocks of Carboniferous age and those of New Red Sandstone times.

We have seen that the effect of the piling up of deposit in Carboniferous times was to produce a generally level surface, and could we have gazed on our area at the close of that period our eyes would no doubt have rested in most places on an extensive flat stretching as far as the eye could reach.

But movements had already begun which were destined to change the aspect of the area in a most marked manner, and these movements went on, no doubt with occasional pauses, during the period of piling up of the New Red Sandstone beds.

As a result of these movements, conditions were produced which in some respects resembled those of Old Red Sandstone times, for the movements were of the mountain-forming type; great ridges stretched across the area, with hollows between, giving rise to hill-ranges bounding flatter tracts. The hill-ranges probably formed a network, as some ran north and south, and others east and west. Thus the Carboniferous rocks of the Pennine Hills of northern England were folded into a ridge which ran north and south, while those of the Mendip Hills were ridged in an east and west direction, and other ridges ran parallel to these two. This network may in itself have produced some effect upon the climatic conditions, though the great change in climate was no doubt due to causes acting over areas reaching far beyond the confines of the British region. What influence this change of climate had upon the characters of the rocks will be presently discussed. At present we are concerned with the effects of the earth-movements on the relation between Carboniferous and New Red Sandstone rocks.

As the general uplift produced continental conditions the area which in Carboniferous times had been receiving deposits of sediment over the greater part of its extent was now subjected to the agents of erosion, and these had naturally most effect on the ridges, which were gradually worn down, while the wreckage produced by their wear was on the whole accumulated in the hollows. As the ridges got worn away materials of later date would be piled over their sites, until they in turn became buried, and in the meantime fresh ridges must be formed by uplift, also to be modified by erosion.

Consequently, we find that the rocks of the New Red Sandstone system rest on those of the Carboniferous system with a great unconformity, and there is a gap between the newest Carboniferous rocks of our country and the lowest New Red Sandstone rocks when erosion was mainly rife over the whole area. This gap indicates a period of long duration, when deposits of importance were being formed in some other parts of the world,—deposits which have no representatives in Britain.

These movements at the end of Carboniferous times gave to our coal-fields the basin-shaped structure which has already been noticed.

Before considering further the conditions of our country in New Red Sandstone times, we must give a fuller account of the characters of the British strata formed in these times.

Characters of the New Red Sandstone Strata. We have stated above that the name which we have adopted for the period was assigned to it on account of the nature and colour of its most abundant deposits. Masses of sandstone varying from pink to deep brick-red form the bulk of the deposits. But there are other strata of importance associated with the sandstones, and the sandstones themselves present features of considerable interest which require further notice.

Conspicuous among the deposits of the earlier part of the period, the Permian strata, we may notice a mass of limestone, known as Magnesian Limestone. This is so called because in addition to the carbonate of lime of which ordinary limestones are formed there is also a

varying quantity of carbonate of magnesia. This limestone stretches from the coast of Durham to near
Nottingham, varying in purity, and having different
thicknesses of sandstones and mudstones associated with
it along its outcrop. The group also contains some
gypsum or sulphate of lime in places. The total thickness
of these Permian rocks is not very great; more than a
few hundred feet of rock seldom occurs.

The upper New Red Sandstone rocks, of Triassic age,
are practically devoid of limestones in Britain; sandstones,
usually red, sometimes yellow, at others mottled with
different tints of red, yellow, and green, make the bulk
of the system, which is often a few thousand feet in
thickness, though the thickness varies greatly in different
localities.

In some districts great masses of angular blocks of
more ancient rocks contribute largely to the strata, with
sand grains in the crevices between the fragments, while
in other places deposits of rock-salt and gypsum are found
associated with the old muds of these Triassic strata.

In other places again we find pebble-beds with re-
markably round pebbles among the sandstones.

The sandstones present other features of interest in
addition to their prevalent red colour. The beds often
thin out rapidly when traced sideways, and the minor
planes of stratification are often arranged obliquely to
the major planes. Again, the individual grains of sand
of some of these accumulations present a noteworthy
appearance. In many beds the grains are unusually small,
polished, and with spherical outlines.

The meaning of these signs will be discovered after
we have noticed the nature of the fossils of the New Red
Sandstone rocks.

New Red Sandstone Fossils. Remains of fossils are common in many of the Carboniferous beds of our area, but in the New Red Sandstone rocks they are generally rare. Only in the Magnesian Limestone are the remains of boneless creatures common, and that only in a few localities, chiefly in Durham. Even when common, the shells of the limestone are often small and distorted, and although individuals may be common, they belong to few forms.

In the other strata of this age fossils are very rare in Britain, and they belong chiefly to bony creatures, especially to amphibians and reptiles, for the true reptiles appear in New Red Sandstone times. Very strange were some of these animals, the heads of some being decked with many pairs of horns. Occasionally the hand-like footprints may be found shewing where they walked across the soft mud when entering or quitting the expanses of water.

The local rarity of fossils of this age is in itself a fact of great import, when we take into account that the rocks of this period in other regions, for instance in the Alps and Himalayas, are often crowded with fossils of great variety.

Conditions under which the New Red Sandstone deposits were formed. Geologists are agreed that the features which we have very shortly noticed may be explained in one way, by supposing that during the period our area was subjected to desert conditions. The British area was part of a continent with a warm and unusually dry climate. In chapter VIII we noticed the effects of erosion in desert regions, and in chapter XI attention was called to some of the characters of the accumulations which are now being piled up in such regions. The reader who has

read what has been written in the earlier parts of the
chapter will probably have recalled some of the descriptions of desert deposits given in chapter XI.

Though the actual effects of erosion, other than the
accumulations, are not often noticeable, they are in rare
cases visible. In Leicestershire a wind-cut mass of rock
has recently been unearthed by the removal of the New
Red Sandstone deposits which were laid over it after the
wind had sculptured it. Here then we have part of a
Triassic landscape restored to us. By permission of
Prof. Watts, who described it, I am enabled to present
my readers with an illustration of this interesting feature
(Fig. 27).

We will now pass to the deposits. The Magnesian
Limestone was clearly laid down in water, but although
the water area must once have had connexion with the
ocean, in order to allow the marine organisms such as
lamp-shells and nautili to gain access to it, it was
probably severed from the ocean when the Magnesian
Limestone was deposited, for the presence of carbonate
of magnesia suggests evaporation, and the fossils bear
resemblance to the creatures of an inland sea like the
Caspian, having no connexion with the ocean. The present
Caspian fauna like that of the Magnesian Limestone is
marked by the little variety of life, and the dwarfed and
distorted characters of some of the shells. Further evaporation of the water tracts is indicated by the deposits
of gypsum, which can only be laid down when a considerable mass of water has been evaporated.

The coarse angular fragments making up a large part
of some of the New Red Sandstone strata recall the
masses of loose material formed at the bases of hill-slopes
in desert regions, while the rounded pebbles and other

Fig. 27. Part of Triassic Landscape.

The wind-worn rock seen on right, owing to removal from it of the Triassic rock which is still seen on left.

deposits may be matched in the beds of the *wadys* of the Egyptian deserts.

The oblique bedding of many of the sands recalls the structure of the sand-dunes of modern deserts, and the small rounded polished grains of these sands are like those of the modern desert sands which were previously described under the name of millet-seed sands.

Late in New Red Sandstone times, lakes were situated here and there among these arid tracts, and the waters of these lakes, like those of modern deserts, received additions during periods of rainfall, which occur even in deserts, and were evaporated during the drier periods. When evaporation took place the salts contained in these lakes were laid down on the floors, to form deposits of gypsum and of rock salt. On the drying muds of these shrinking lakes creatures impressed their footprints. Occasional showers of rain left little pittings on the mud, and the drying mud shrinking in the heat was cracked in different directions, some of these cracks extending for many yards in length, and being of considerable depth, and sometimes having a width of one or two inches.

In these conditions it is not likely that many remains of organisms should be preserved, for in desert regions land and water tracts alike are not favourable to the existence of many plants and animals, and accordingly the scarcity of fossils in New Red Sandstone rocks comes as a confirmation of the conclusion which we have reached from other evidence, that in New Red Sandstone times the British area was a barren desert.

Perhaps in the whole of the sediments of our country no change is more striking than that between the Carboniferous and New Red Sandstone rock-groups, the one indicative of a marshy plain with abundant plant growth due to a humid climate, the other pointing to tracts of upland and lowland in the interior of a continent with a dry climate and scanty signs of life.

Scenery of the New Red Sandstone Rocks. As the rocks of this age are mostly worn away somewhat readily by agents of erosion, the tracts occupied by them are on the whole marked by fairly flat ground. The Permian rocks it is true give rise to fairly high undulating country in places, especially where the Magnesian Limestone appears at the surface, but the Triassic sandstones and mudstones are found in a tract which has few elevations. The rocks of this age extend from the mouth of the Tees through Yorkshire into the Midland Counties, where they extend over a considerable area, and are continued southward from thence to the neighbourhood of Exeter, while another tract extends from the Midlands in a north-westerly direction to the estuaries of the Dee and Mersey and parts of south Lancashire, and little patches are formed in the Vale of Clwyd in North Wales and that of Eden in Cumberland and Westmorland with an extension to the Cumbrian coast; these tracts are now chiefly lowlands usually flanked on either side by higher ground.

In addition to the prevalent lowness of the country we may note the influence of the colours of the sandstone. The landscape often has a rich red hue, which is particularly pleasing when the sandstone rises as a cliff above a river, perhaps topped with a group of pines, or where the sea has eroded a cliff in a set of beds somewhat harder than usual, as at St Bees' Head on the Cumbrian coast.

Lastly, as the soil produced by the sandstone is often fertile, we usually find the ground occupied by rocks of this age in Britain marked by a high state of cultivation.

CHAPTER XXIII.

THE SECONDARY ROCKS.

2. THE JURASSIC ROCKS.

AT the close of Triassic times a marked change came once more over the conditions of the British area, and about four thousand feet of sediments belonging to the Jurassic system were laid down mainly in the open sea, though land appears at no time to have been far distant, and the sediments are therefore chiefly of a type pointing to shallow water.

To allow of the formation of these marine sediments it is clear that the old land-surface on which many, at any rate, of the Triassic rocks of our area were formed, sank beneath the sea, and the Jurassic period was, on the whole, one during which the sea-floor so produced continued to sink, though the piling up of sediment generally kept pace with the sinking of the floor, so that the waters of this portion of the Jurassic ocean never became very deep.

The Jurassic rocks of our isles are most widely spread in England, though patches also occur in Scotland and Ireland. In England they form a continuous strip of variable width which stretches from the coast of north Yorkshire, through Lincolnshire, and then in a south-westerly direction to the shores of the English Channel in Dorsetshire.

The trend of this strip is due to movements which occurred long after the deposit of the rocks, and caused them to be tilted with a dip towards the south-east, so that they plunge beneath newer rocks to the south-east, and must once have extended from their present limits in a north-westerly direction, but most of the rocks of this age to the north-west of the strip have been since removed by erosion.

It may be here observed that the general trend of the New Red Sandstone rocks below the Jurassic strata, and also that of most of the newer sediments of England, is due to the same set of movements, the nature of which will be more fully discussed in a later chapter.

Of the sediments which compose the Jurassic strata of Britain the greatest thickness is composed of clays, which form considerably more than one-half of the entire group. These clays are largely found in three great bands, separated from each other by other kinds of deposit. The lowest band is known as the Lias. Some writers divide the Jurassic rocks into Lias below and Oolites above. Above the Lias we find, in the south-west of England especially, some of the oolitic rocks, placed between the Lias and another great mass of clay which overlies them. This group of oolitic rocks may be called the group of Bath Oolites, as it is well known around Bath. Its sediments are variable, but some important bands of limestone, having what is known as oolitic structure, occur in them. This oolitic structure causes the rock to exhibit small grains looking somewhat like fish-roe, the grains being about the size of pins' heads. Into their mode of production we will not here enter.

In Yorkshire the Bath group of Oolites is largely represented by a great group of sandstones formed in estuaries and containing among the sandy bed some thin seams of coal.

The important mass of clay above these Oolites is known as the Oxford clay, as it is well developed around the city of Oxford, though it is traceable along the whole line of the strip occupied by the Jurassic rocks.

This clay is in turn covered by variable sediments, of which one of the most interesting is a limestone formed largely of reef-corals in places, and accordingly this is known as the Corallian division.

Another important clay succeeds. It is called the Kimeridge clay, from a village of that name in Wiltshire, though it also is continuous along the strip.

The uppermost division of the Jurassic rocks forms the Portland group of Oolites, being well seen in the island of Portland. The rocks of this group are absent in central England, but are found in the southern counties as far north as Oxfordshire and again in Yorkshire.

The deposits of this group are again variable. In Portland an important limestone occurs; but one of the most interesting sets of sediment lies above this limestone, and to it is applied the term Purbeck series, from its occurrence in the Isle of Purbeck. In that district it is formed of estuarine and terrestrial beds, of great interest from the nature of the fossils, which will be presently considered.

The Jurassic Fossils. Owing to the great gap which occurs in Britain between the Carboniferous and New Red Sandstone rocks, and to the paucity of fossils in the latter group of rocks, the contrast between the Carboniferous and Jurassic fossils of our country is very great.

The plants of the Jurassic rocks differ in many respects from those of the Carboniferous rocks, but we cannot enter into an account of these differences.   The trilobites, those creatures which form so interesting a feature in the life of Primary times, have completely disappeared.   The lamp-shells which were so abundant in the Primary beds are much rarer, and the proportion of true molluscs is now much higher.   Among these true molluscs, the shells of certain creatures closely related to the pearly Nautilus are very abundant in the rocks of the Jurassic system and also in those of the system which succeeds it, and the same is the case with certain allies of the living cuttle-fishes.

Fig. 28. An Ammonite.

The former creatures are known as Ammonites, and the latter as Belemnites.

The shell of an Ammonite which is represented in Fig. 28 is not unlike that of the Nautilus, but there are

differences in the internal structure which at once enables
us to separate them; and each is readily distinguished
from such shells as those of the whelk and periwinkle by
the fact that the inside of the shell is separated into
chambers by regular partitions, the edges of which are
sharply folded in the case of the Ammonite shell, so that
when the shell is stripped off, and a cast of the inside is
seen, the edges of these partitions appear in crinkled
lines, a feature shewn in the figure which represents a cast
of the interior of the Ammonite, formed of hardened mud,
the shell itself having been removed.

Fig. 29. A Belemnite.

The part of the Belemnite which is commonly preserved is a solid, limy, dart-shaped body, usually of a
brown colour, shewn resting on the shelf in Fig. 29.
These fossils have been fancifully spoken of as fossil

thunderbolts, with which of course they have nothing to do.

But the remains of creatures which possessed bones are among the most interesting and remarkable fossils of the Jurassic period. Many of the fishes are of strange forms, but to these we will not further allude. The reptiles of the period are abundant, and many must have been indeed of weird aspect. So abundant are the remains of these creatures in some beds, and so strange their forms, that the Jurassic period has been well spoken of as the Age of Reptiles.

Some of these creatures swam in the shallow seas, darting here and there and feeding on the fishes and on smaller reptiles. In many cases their legs were planned to act like paddles, enabling them to swim rapidly, and the tails of some ended in a huge fin. The creatures of one group possessed extraordinarily long necks, and have been fancifully compared to a snake threaded through the body of a turtle. Those of another group possessed bat-like wings, and flew through the air.

Birds, too, have been found in the Jurassic rocks, though hitherto their remains have not been extracted from any British beds. These birds possessed long lizard-like tails with rows of feathers along the tail, and the jaws were provided with teeth!

The highest of the backboned animals, the mammals, are also found in these strata. Jaws of small mammals have been discovered in an estuarine deposit in Oxfordshire and others in an old land-soil of the Isle of Purbeck. These early mammals, which are not quite the earliest, as a few also occur for the first time in the Triassic rocks, belong to the most lowly groups, their living allies being the duck-billed platypus of Australia.

Conditions under which the Jurassic rocks were deposited. It has been already said that the deposits are of shallow-water character. Land cannot have been far away. There is evidence that a great continent lay over the arctic regions, coming down to the north of Scotland, and during part of the period even further south. Other masses of land must have occupied other parts of the British area, one certainly lying under London and the neighbouring parts during a portion of the period, and others elsewhere. In the shallow seas between what was probably a group of islands the sands, muds and limestones of the period were laid down. Sometimes the sea was clear in parts of our area, allowing of the formation of limestones by the remains of reef-building corals and many other creatures. At other times rivers poured in their volumes of sand and mud, and the turbid seas were then rendered unfavourable to the growth of reef-building corals. In the estuaries and along the coasts the great sea-reptiles darted about in search of prey, while overhead wheeled the flying reptiles. On the land grew the plants of the period and the small mammals wandered, their corpses being embedded in the soil, or washed down by the rivers to be embedded in the sediments of the estuaries.

The climate was almost certainly warm, as indicated by the frequent occurrence of coral reefs.

Altogether the conditions of the period must have resembled some parts of the coast of Australia and the islands of the coral sea in its neighbourhood at the present time.

Scenery of the area occupied by Jurassic rocks. The soft clays are readily eroded, and are chiefly marked by plains. The plain of the Lias is continuous with that of the Trias, which borders it to the north and west. The Oxford and Kimeridge clays give rise to other plains, which owing to the absence of large masses of hard rock between them in Cambridgeshire and Lincolnshire have caused the formation of the Fenland of that district. The harder rocks of the Bath group of Oolites tend to stand out, as they resist erosion, and are often marked by hilly ground, which runs in long strips parallel to the strike of the rocks, the limestones especially forming terraced hills with their steeper slopes facing north-westward, and their gentler slopes falling south-eastward in the direction of the dip of the rocks. In places the Corallian rocks give rise to similar features on a smaller scale. The soils furnished by the weathering of many of the harder beds are often fertile, and much of the country occupied by these rocks is extensively cultivated, the plains occupied by clays being chiefly marked by grass, except in the Fenland where a more modern accumulation of peat which covers the Jurassic rocks has rendered the soil very fertile.

CHAPTER XXIV.

THE SECONDARY ROCKS.

3. THE CRETACEOUS ROCKS.

THE rocks which come next in order are called *Cretaceous* because among them is a thick mass of sediment composed of *chalk*. These rocks lie to the east and south-east of the Jurassic rocks in England, and patches are also found in Scotland and Ireland.

In England they extend from the Yorkshire coast through part of east Yorkshire, Lincolnshire, some of the East Anglian counties, and then, in a south-west direction to Hampshire and the Isle of Wight, and from this strip another extends in an easterly direction, through Surrey, Sussex, and Kent, to the coast between the mouth of the Thames and the district around Brighton.

The lower strata of the Cretaceous system consist of varying thicknesses of sand and clay. In Kent and Sussex where the Hastings Sands at the base are succeeded by Weald Clay, and this by Lower Greensand, these beds are, with the exception of the topmost strata, of freshwater origin, while in Yorkshire they are marine.

Above them is a mass of clay in the south-eastern counties as far north as Norfolk. This clay is known as Gault, and in northern Norfolk, Lincolnshire, and Yorkshire it is replaced by a thin deposit of limy rock known as the Red Chalk.

In some of the southern counties a curious deposit containing little greenish coloured grains is spoken of as the Upper Greensand. Some at any rate of the Upper Greensand is of the same general age as the Gault.

The Gault and the other deposits formed at the same time are of marine origin, and were laid down in waters of varying depth.

Lying over the Gault and its representatives is that most extensive of our Cretaceous rocks, the Chalk, which with local variations retains the same general characters over the whole of that part of England where it exists.

It is usually a white, earthy limestone with many layers of flint running parallel to the bedding-planes in its upper portion. The greatest thickness of Chalk in England is found in Norfolk, where over eleven hundred feet of its strata have been proved to exist.

When looked at with the naked eye chalk appears to be formed of very small particles, but if it be examined under a microscope after proper preparation, it is found to consist very largely of the perfect or broken remains of fossils of various kinds. Fragments of shells, sea-urchins and other creatures occur plentifully, but much of it is largely composed of the remains of those foraminifers which we spoke of in chapter XII, adding largely to some of the modern sediments which are now being formed in the open oceans.

The flints have been formed since the Chalk was deposited by the collection around definite points of silica which was once scattered through the rock. We have reason to know that this silica originally existed in the chalk owing to the presence of sponges and other creatures which secreted silica in the Chalk sea.

The planes of stratification in the Chalk are often well seen, and the rock is now found in many cases to be traversed by very regular joints, which are of importance from the way in which the erosion of the Chalk is partly affected by them.

Fossils of the Cretaceous System. The fossils of the Cretaceous rocks bear a general resemblance to those of Jurassic times. The plants of the Lower Cretaceous rocks resemble those of the Jurassic beds, and although a great change in the nature of the vegetation took place when the Upper Cretaceous rocks were laid down, plant remains of this age are very rare in Britain. Among the animals Ammonites and Belemnites were still prominent, and the creatures having bones were of the same general types as those found in the Jurassic rocks.

In the freshwater beds of the Lower Cretaceous strata of southern England are embedded remains of freshwater shells very similar to those now living in our rivers.

In the Chalk, the fossils are remains of animals which as a whole resemble those living in the open oceans at some distance from the land, either swimming or floating on the surface-waters, or inhabiting the ocean-floors. Among the shore-living animals of modern times that group of molluscs which form shells of one valve such as the whelk and periwinkle are rare.

It has often been found that prior to the final disappearance of a group of animals they undergo very remarkable changes of form or size, and this is the case with the group of ammonites. An ordinary ammonite has a shell which is coiled into a flat spiral form, but the shells of this group in the Chalk, besides being of this form, also assume many other shapes. Some are straight, others curved and hooked, others again are partly spiral and then become straight, and finally hooked back, while others are coiled into a spiral of which the coils do not lie in one plane. Otherwise each shell possesses those structures which characterise the ammonite group.

Conditions under which the Cretaceous rocks were deposited. The Lower Cretaceous rocks of our country were formed under conditions not differing in any marked degree from those which existed during the formation of the Upper Jurassic rocks. The lowest Cretaceous beds of the south-eastern counties resemble those forming the Purbeck strata, while those of Yorkshire are like the uppermost Jurassic rocks of that county. In the counties to the north and north-west of London, land existed during the formation of most of the lower Cretaceous rocks elsewhere, and higher rocks of this system there repose unconformably upon the Jurassic beds.

As the area sank marine conditions became widespread through extensive tracts of England, and the Gault was laid down in a sea which was not too remote from the coast-lines to allow of the transport of mud from the adjoining lands in some quantity, save where the deposits of the period are represented by the Red Chalk.

At last the tract sank to such an extent that the coast-lines receded to so great a distance that little earthy sediment was carried to the area, and then the Chalk began to form upon the floor of the open sea.

A detailed study of the nature of the Chalk, a comparison of its strata with those of modern seas, and observation of the places where the Chalk disappears on the Continent and is replaced by masses of clay and sandstone, suggest however that although the southern British area was fairly remote from land when the chalk was formed in it, the sea-tract was nevertheless not part of an open ocean, but rather resembled a great gulf surrounded on three sides by the land. To the north lay that arctic continent which existed also in Jurassic times, to the east the nature of the Cretaceous rocks of Hanover, Saxony, and Bohemia, suggests the existence of land not far to the east of those countries, while to the south there is evidence of land extending through central France and separating the sea in which the Cretaceous rocks of the Pyrenees and Alps were laid down, from the chalk gulf of the north.

The conditions in fact probably resembled those which now exist in the Caribbean sea, in the deeper parts of which sediment is now being piled up which in future times may be converted into a rock not unlike our own Chalk.

With regard to the climatic conditions of the period in Britain, we have not much evidence, though the nature of the molluscs and other organisms suggests a fairly warm climate during a considerable part of the period

Scenery of the Cretaceous rocks. The clays of the older Cretaceous rocks are usually marked by low-lying ground like part of the Weald of Kent and Surrey, while the sandstones are often marked by hilly tracts, which is also the case with some of the country south of London which is occupied by the Upper Greensand.

The Gault again is usually marked by flat country.

The scenery of the Chalk country is very characteristic both as regards the inland tracts occupied by this rock and the coast. Inland the chalk forms downs, undulating hilly expanses, which, owing to the porous nature of the rock, have few streams coursing over the surface, and are covered with short sweet herbage, often made bright by the abundance of flowers of such plants as are adapted to the dry surface.

On the coasts, owing to the well-jointed nature of the rock, imposing cliffs like those of Yorkshire about Flamborough Head, of Kent between the North Foreland and Dover, of Sussex at and near Beachy Head, and of the Isle of Wight, occur. These are "the white cliffs of Albion," so familiar to those of all times who have landed in Britain from the opposite shores of France. Not only are the cliffs high and precipitous, but the sea often wears arches through projecting headlands, and eventually the ends of the headlands may be severed from the mainland to form isolated stacks and needles—outstanding columns of chalk of which the well-known Needles to the west of the Isle of Wight furnish typical examples.

CHAPTER XXV.

THE TERTIARY ROCKS.

THE EOCENE ROCKS.

AT the close of Cretaceous times marked changes once more took place in the physical conditions of our area.

The sinking of the sea-floor on which the Chalk was laid down ceased, and was succeeded by uplift causing the rise of much of the area above sea-level, and although parts were afterwards sunk once more beneath the waters to allow of the formation of the Eocene rocks, these water areas were much more restricted than those of Cretaceous times, and the rocks of Eocene age are not only found over a far more limited area but their characters are very different from those of the Chalk.

The British Eocene rocks are of two very different types. In the south of England, shallow-water sands and muds, with occasional thin limestones are found in two areas, once continuous, but now separated owing to the occurrence of uplift and erosion after their deposition. These two areas occur, the one around London, the other in southern Hampshire and the northern part of the Isle of Wight. As the result of the movement to which we have just alluded was the folding of the strata into shallow troughs, these two areas are spoken of respectively as the London and Hampshire Basins.

The other area occurs in some of the Western Isles of Scotland, as Mull, Skye, and Staffa, and in Antrim in north-east Ireland. Very different is the nature of the rocks of that area from those of the London and Hampshire Basins. Instead of the soft sands and clays which form the chief parts of the deposits of the south, we here meet with great sheets of basalt, with volcanic ashes, occasional thin clays and sands, and sometimes a peat-like mass largely composed of the remains of plants, the whole pointing to successive outpourings of volcanic materials upon a land-surface.

In the London Basin, the lowermost Eocene beds consist of variable sands and muds, which are partly freshwater and partly marine. They are of no great thickness. Above them is the chief deposit of this basin, known as the London Clay. It is about five hundred feet thick, and consists of a bluish clay. The highest beds of the basin are sands of no great interest.

The lower beds of the Hampshire Basin resemble in general respects those around London, but higher beds are here found, consisting of sands, clays, and some thin limestones, some of the strata being of marine origin, while others were formed in fresh water.

The rocks of the Scotch Islands and of north-eastern Ireland are in many ways of interest. The most remarkable of these rocks are the great masses of basalt which occur in nearly level sheets. These basalts were poured out upon the surface as lavas, often sealing up the terrestrial deposits which were formed in the intervals between the outpouring of successive flows.

Fossils of the Eocene beds. The chief fossils of the

land-deposits which are sealed up between the lavas of the island of Mull in Scotland are plants, and similar plants are found in some of the beds of the London and Hampshire Basins. They are nearly allied to many existing forms, though differing from those which grow wild in our own country at the present time.

Fig. 30. Rock with Nummulites.

The animals devoid of bones are also very like living forms. Corals, sea-urchins, crabs, and molluscs extracted from these beds have a familiar appearance. The actual

forms are in most cases different from existing ones, but
they are very nearly allied. In some beds in Hampshire
some curious coin-like bodies about half-an-inch in
diameter are preserved. They are the hard parts of
foraminifers of a particular kind, called *Nummulites* on
account of their coin-like appearance (see Fig. 30). These
fossils largely build up great masses of limestone in other
parts of Europe as in the Alps, and in Asia as in the
Himalayas. Similar limestones are found in Northern
Africa, and the Pyramids of Egypt are built of a limestone
rock which is crowded with these Nummulites. The
Nummulites are fossils which when found in abundance
indicate that the containing rocks are of Tertiary age,
and we therefore give a figure shewing their appearance.
The reader will easily remember three very typical fossils
of the three great groups of rock: the Primary rocks
contain Trilobites, the Secondary rocks Ammonites, and
the Tertiary rocks Nummulites.

The Eocene back-boned animals of our country are of
interest, though we have not so great a variety as is
furnished by the rocks of some other countries, especially
those of North America. Of the fishes various sharks
lived in the Eocene waters of our area. Of reptiles we
find among others the remains of sea-snakes and of large
turtles. A few birds have been discovered, and a greater
number of mammals. The latter are mostly of much
more advanced types than the simple mammals of the
Secondary rocks. Some were of very remarkable appearance, and in many respects unlike any mammals now
living.

Conditions under which the Eocene rocks were deposited.
It was remarked that many of the Eocene beds of the
London and Hampshire Basins were of freshwater origin,

and that those of marine origin were formed at no great distance from a shore-line. The actual characters of the sediments prove this to some extent, but it is particularly shewn by the nature of the included fossils. Those of the London Clay may be taken as an example of the light which the fossils throw on the conditions under which the beds were deposited.

The molluscs, crustacea, and other creatures are all allied to forms which at the present day live in shallow waters. Some of the bones are those of creatures which do not live far from the coast. The abundance of plant-remains in parts of the clay shew that a large river must have been near, which brought these plant-relics down to the sea. The London Clay then appears to have been deposited in a tract of the nature of a large and wide estuary.

These fossils also throw some light upon conditions of climate at the time of formation of the London Clay. We have already noted that this kind of evidence is not always easy to apply. Creatures are able to adapt themselves to different climatic conditions. The living tiger, usually found in warm regions, has been found wandering not far below the snow-line in the Himalayas, and we shall presently see that elephants, somewhat different it is true from those which now exist, lived in countries which were subjected to an arctic climate. We have therefore to be very careful in arguing as to climatic conditions after examining only a few forms. But in the London Clay plants as well as animals tell the same story. Of the animals, those which lived on land point to the same conclusion as those which dwelt in the water, and the boneless animals lead us to the same conclusion as do those with bones. All the fossils point to the occurrence of a warm climate in our area during the deposition of the London Clay. The plants give particularly strong evidence. Remains of palms and other plants are very similar to the corresponding parts of plants which now live just outside the tropics.

We have other evidence. In passing north from Britain we find that the plants of Eocene age indicate that, though a colder climate than that of Britain marked more northerly latitudes, nevertheless, the climate was warmer than that which exists in the particular locality at the present day. Thus in Greenland Eocene plants are found in beds occurring in a region which is now nearly barren.

Again, as we study the newer Tertiary rocks, the fossils shew signs of a gradual lowering of the temperature, which as we shall presently see led in late Tertiary times to the occurrence of arctic conditions in Britain.

We need have little hesitation therefore in accepting the evidence of the fossils of Eocene times, closely related as they are to various existing forms of life, as indicative of the occurrence of a warm climate in Britain during the period.

Scenery of the Eocene tracts. The soft, slightly consolidated rocks of the London and Hampshire Basins are somewhat readily eroded, and the ground is usually low, though undulating, but the sands of the highest Eocene beds of the London Basin are often found capping hills, as those of Highgate, Hampstead, and Harrow near London, those near Brentwood in Essex, and those of a considerable tract of country south of Reading, much of which is covered by a growth of firs which these sands readily support.

The volcanic rocks of Scotland give rise to a grander

type of scenery. The basalts resist erosion to some extent, and rise into high ground, often terraced, owing to the alternation of the basalts with more easily eroded rocks. Great masses of plutonic rocks were forced through these basalts, and, having resisted erosion, stand up in high hills, as in Mull and Skye. The erosion of the well-jointed basalts by the sea has given rise to high sea-cliffs in some of the Scottish Isles and on the Antrim coast, and the curious jointed structure which basalts sometimes possess, forming a series of many-sided columns, causes the remarkable appearance which these basalts present in some places, of which the best known are in the island of Staffa, and at the Giant's Causeway.

CHAPTER XXVI.

THE TERTIARY ROCKS.

2. MIOCENE TIMES.

THROUGH Secondary times, and to some extent in Eocene times, large tracts of our islands were under water and received the deposits of sediment which have been briefly described in the preceding chapters. It is true that every now and then uplifts of portions of the area occurred which were marked by erosion, but these long periods were on the whole characterised by accumulation of sediments.

In Miocene times changes of great importance occurred giving rise to continental tracts. These changes, as far as our area is concerned, are to some extent comparable with those which took place in Devonian and New Red Sandstone times respectively. Not only were the British Isles affected, but also wide tracts of the continents of Europe and Asia. There was a general uplift turning the Eocene seas into lands save in a few tracts of small extent, and in addition to this building of continents, great earth wrinkles produced important mountain ranges such as those of the Himalayas and the Alps, for we find Eocene strata in a greatly folded condition, thousands of feet above sea-level in those mountains, while beds of the period following the Miocene, namely, the Pliocene period, rest nearly undisturbed against the bases of the mountains, which shews that the main uplift was in Miocene times.

As one consequence of this uplift there is, as far as we know, a complete absence of Miocene rocks in Britain. The continents and oceans were becoming similar as regards general distribution to what they are at present, and accordingly, apart from the few insignificant patches of Miocene age which are found on the Continent, we must seek for the marine strata of this period beneath the waters of the existing oceans.

But though Britain presents us with no rocks of Miocene age, the changes which occurred in the Miocene period are by no means without their effect on British geology. How far the northern and western parts of our area existed as land at the end of the Eocene period is a question not yet settled, but that part of the south and east tracts were occupied by water has been shewn in the last chapter.

As the result of the continent-building movements of Miocene times this portion was now formed into land, and beds which occupied that tract were tilted and eroded. This tilt in Yorkshire and Lincolnshire and the land immediately to the west of those counties caused the strata to be raised on the west and sunk in the east, thus giving them an easterly dip. Further south the uplift produced a south-easterly dip, and in the extreme south, owing to the production of the great ridge which separates the Hampshire and London Basins of Eocene rocks, and of minor ridges of which part exists in the south of the Isle of Wight, the dip of the strata is sometimes northerly, at other times southerly. This is shewn in

Fig. 31 which is a section of the strata from the north of the London Basin to the south of that of Hampshire. The occurrence of Pliocene beds at P. which will be again noticed, is of importance as fixing the date of the uplift. The Eocene beds once ran across the arch separating the two basins. Uplift was necessary to allow of the erosion of these Eocene beds, and as the Pliocene beds here rest on Cretaceous rocks it is clear that the uplift which allowed of this erosion was before the Pliocene period, in other words, it was in Miocene times.

The eastern and southern part of the country then has existed as land from the Miocene period. All the rivers which discharge their waters on the eastern and southern coasts between the mouth of the Tees and that of the Exe took their origin in Miocene times and such erosion as they have produced dates from that period. It is interesting to know that the surface features of at any rate a considerable tract of England are thus from the point of view of the geologist very modern.

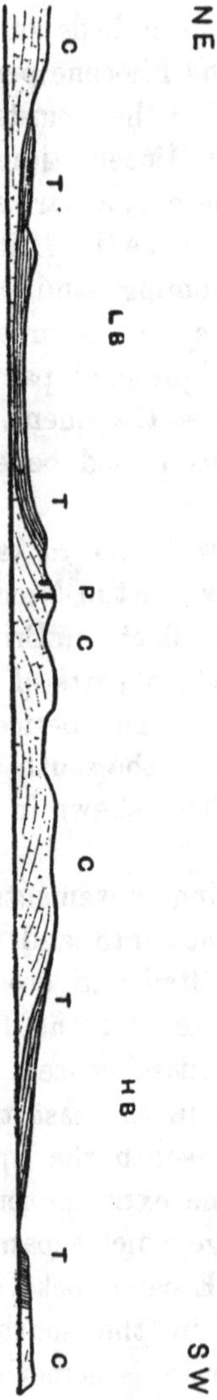

Fig. 31.   Section across the London and Hampshire Basins.

LB.   London Basin.   T.   Tertiary beds.
HB.   Hampshire Basin.   C.   Chalk.
P.   Pliocene beds in 'pipes.'

3. The Pliocene Beds.

In Pliocene times, parts of our country were again submerged, but these tracts lay close to the present coast-line, and it is there that we meet with beds of the Pliocene period. The principal strata of this age occur in Norfolk and Suffolk, but deposits of some interest are found in Kent and also in the south-west of Cornwall.

The beds of Norfolk and Suffolk are chiefly of the nature of shelly sands, locally known as 'Crags.' These crags are of no great thickness, and were chiefly formed as sandstones, though one of them may have been laid down at some little distance from the coast. They are of slight thickness, and are found in somewhat scattered patches.

The Pliocene beds of Kent are very local. They exist in hollows worn in the chalk by solution, and have been preserved from the erosion which has removed the main mass. These patches occur on the North Downs, and are chiefly of interest on account of the evidence which they supply, as already noted, that the uplift of the arch into which the beds of our south-eastern counties were thrown took place before the Pliocene period.

The Cornish beds which are found near the Land's End are also shelly sands, and we need not here notice them further.

Fossils of the Pliocene Beds. The creatures which do not possess back-bones are largely molluscs, though other creatures occur. The greater number of these belong to forms which are still existing. In many of the Crags we find shells of the common cockle, mussel, oyster, whelk, and periwinkle, but a few of the Pliocene forms of mollusc are now extinct.

The back-boned animals are largely extinct but they are very closely related to living forms.

One of the features of interest of the Pliocene fossils is that they shew a gradual lowering of climate during the progress of the Pliocene period. In the earliest Crags some of the shells are forms which are now found in the seas to the south of our area. In the middle of the period the fossils indicate that the temperature was much as at present, while at the end of the period we find a number of forms which now only live in the seas to the north of the British Isles.

The very uppermost bed of Pliocene age at Cromer contains plants. These are the arctic birch and arctic willow, and we here get the first indications of those severe climatic conditions which prevailed over the British Isles during the next period.

Conditions under which the Pliocene rocks were deposited. The climatic conditions have just been considered. As regards other conditions, we have evidence that the Straits of Dover did not exist, but that a barrier of land separated the Pliocene English Channel from the water area which occupied part of what is now the North Sea and the strip of country which forms the east of Norfolk and Suffolk. In this water area the East Anglian Crags were laid down as deposits of an estuarine character, for in some of the Crags we find the remains of freshwater molluscs and even of land molluscs associated with those of marine habit. The estuary was mainly that of the Pliocene Rhine.

As the Pliocene beds have so limited a distribution and are of no great thickness they produce little effect upon the scenery of the country.

CHAPTER XXVII.

THE TERTIARY ROCKS.

4. THE PLEISTOCENE PERIOD.

A. *The Great Ice Age.*

AT a time which compared with the long periods which have elapsed since the formation of the Precambrian rocks is but as yesterday, two events of singular interest occurred in our area. The one was the occupation of large tracts of our island by great masses of snow and ice, the other the first appearance of man.

Each of these events being of so great interest, it will be convenient to treat of the two in separate chapters.

We have previously (in chapter IX) gained some acquaintance with the action of moving ice, and it now remains for us to apply our knowledge to the study of the effects of ice as shewn in our own country.

The date of the glacial deposits is shewn in Norfolk along the coast at Cromer, where the uppermost Pliocene deposits are overlain by great masses of *boulder-clay*;— clay containing angular stones which are blunted, and often scratched and polished, many of which can be proved to have been borne from far distant regions, some indeed having come from the peninsula of Scandinavia.

Now though there is no doubt that these boulder-clays of Norfolk were formed by ice, there is a difference of opinion as to the exact method of their production, and therefore, though we will presently give some idea of the distribution of these glacial deposits which are widely spread over the lowland tracts of Britain, it will be well in discussing the actual nature of the ice-work in Britain to pay attention mainly to the features displayed in our upland region, as to the origin of which there is no doubt.

In visiting one of these upland regions the most obvious effects of glaciation are obtainable by noticing the marks of ice on the solid rocks over which that ice has passed in its course along the sides and floors of the valleys.

In the ninth chapter the chief changes which were produced in such rocks by ice were noted, and among other things it was seen that the grinding action of the ice was most noticeable on those sides of projecting rock-masses which faced the direction from which the ice was coming. Accordingly, if we visit some upland glen of Britain as the Pass of Llanberis in North Wales, Borrow-dale in Cumberland or Glencoe in Scotland, when looking up the valley we may see little or no effects of ice. On walking some distance up the valley and looking *downward*, however, we shall probably be struck by the rounded appearance of the projecting rocks of the sides and floors of the valleys, which form a marked contrast with the rough and often bristling masses of rocks which crown the ridges between valley and valley. A further inspection will often shew that these rocks are polished to a degree greater than that shewn by water-worn stones; and if we look about, and especially if we chance upon a surface of fairly hard fine-grained rock from which the turf has been recently removed, we shall probably find some of those parallel flutings and scratches which we have already found to be typical of ice-action. Further study convinces us that in addition to the grinding action produced by glaciers, these moving ice-masses have also torn away masses of considerable size from the parent rocks. It is held by many geologists that, owing largely to such tearing processes, hollows of considerable size may be scooped out in the track of the ice, which, when the ice retreats, become filled with water, to form lakes.

The existence of glaciers in our country during the Great Ice Age is shewn not only by the marks of erosion, but by the occurrence of the deposited materials which were transported by the ice on its surface, in its interior or at its base.

In the first place we often notice large blocks of rock which can be shewn to have been brought from some distance higher up the valleys than their present resting-places. These blocks are often perched on the summits of cliffs, in such a way that they must have been gently stranded on the spot on which they now rest. Had they rolled from the cliffs above they would have been in many cases carried over the cliffs on the tops of which they are actually found.

At the present day such blocks are often stranded in cold regions by glaciers and also by icebergs. There is evidence that in the case of the perched blocks of many of our upland valleys, glaciers and not icebergs were the agents employed in their transport.

In some cases long trains of blocks are found on the valley sides, resembling in all respects the lateral moraines formed by existing glaciers. These lateral moraines are not very common in our country, as they tend to be destroyed by erosion, or covered by other accumulations after their formation.

More common are the mounds of ice-carried materials which formed the terminal moraines of the glaciers. In many of the upland valleys of our country these moraines occur as groups of mounds often arranged in crescents, of which the convex side faces down the valley, each crescent being separated from the next by more level ground. Such crescentic moraines mark periods of pause, during which the end of the glacier stayed at the same place for some time, while the flatter tracts between the crescents are due to the retreat of the glacier, which took place quickly enough to prevent the piling up of much material at its end in any one spot.

These moraines have often formed barriers across the valley, resembling the dams made by man to form ponds, and many lakes are due to these moraine dams, while others may be due partly to erosion and partly to the formation of barriers and moraine stuff.

In passing from these upland regions to the flatter tracts of our country the evidences of past glaciation become in some respects more striking.

Extensive areas are covered, often to a great depth with masses of boulder-clay, which occur not merely in the lower parts of the valleys, but often on the ridges and plateaux. This boulder-clay is usually a very stiff clay with no trace of bedding, and crowded with boulders of various sizes, some being occasionally as big as a house. The tracing of these boulders to their parent rocks has thrown much light upon the direction of their transport, and many of them are traceable to the rocks of our upland regions. In many parts of the Midlands boulders which have been derived from North Wales are common. In parts of Lancashire, Cheshire and other adjoining counties, as also in South Durham and North Yorkshire, boulders are found which have come from the Lake district, and in east Yorkshire and as already observed, in Norfolk as near Cromer, some of the boulders are of Scandinavian origin.

In the Isle of Man and in the north of Cumberland and at other places we find boulders of Scotch rocks.

Now without entering into a discussion as to the exact mode of transport of all these boulders, we may state that the evidence which they furnish goes to shew that at the time of their transport the upland regions were glaciated in a manner rather recalling the present glaciation of Greenland than that of Switzerland. Hill and valley alike must have been buried beneath snow and ice, and these great ice-sheets spread beyond the limits of the upland region, though it by no means follows that every tract which is now occupied by boulder-clay was covered by land-ice, for at the present day boulders are carried far beyond the outer limits of land-ice by water action.

This boulder-clay occupies many tracts of Britain, north of a line drawn from the mouth of the Severn to that of the Thames. South of that line it is doubtful whether any true boulder-clay was formed.

We gather therefore that in one way or another the greater part of the British Isles were affected by glacial action during the Great Ice Age.

The glacial relics of our upland regions which have been briefly described above were mainly produced by the valley glaciers which lingered in those upland regions after the period of greatest extension of the ice.

The glacial period was no doubt short as compared

with such periods as those when the Jurassic and Cretaceous rocks were formed. We can get no estimate in years of the duration of the period, or indeed of the length of time which has passed since the period ended, though the latter was certainly great as compared with the duration of historic times.

We are still much in the dark as to the causes which brought about the Great Ice Age. They were widespread, for at the time when our country was glaciated, other countries were also affected by glacial action. Large tracts of northern Europe and of the northern part of North America give proofs of glaciation, and the ice of the mountainous tracts of more northerly areas also extended far beyond the present limits, as proved in the case of the Alps and Pyrenees, while other hill-districts which now contain no glaciers once possessed them, for instance the Harz Mountains in Germany, and Lebanon in Asiatic Turkey.

These causes of glaciation, whatever they were, at last ceased to operate in our area and some other countries. The climate improved, and the masses of ice gradually retreated. No doubt long after the glaciers had shrunk away from the lower ground little glaciers would linger in the high level combes or corries,—those small, half-bowl shaped valleys which are so frequently found beneath our higher mountain tops. As the influence of ice ceased in the lowlands other conditions became of importance, and a set of deposits, also referred by most geologists to the Pleistocene period, began to be formed, which we shall consider in the next chapter.

The glacial deposits of our country do contain fossils, though rarely. Some of these have been carried by ice from elsewhere, and sometimes belong to a period before the Ice age. The boulder-clay of Yorkshire and East Anglia among other places often yields abundance of shells, usually in fragments, which have clearly been carried from a distance, and some of them give signs of having been derived from Pliocene beds.

Such fossils as can be shewn to be of Glacial age naturally give proofs of cold conditions; the shells for instance being related to those which now live in the Arctic seas.

Some of the effects of the glaciation of our country upon the scenery have already been noticed. We need not again allude to the grinding and tearing of rocks, the piling up of moraines, and the formation of some lakes. The boulder-clay of the lowland regions is often responsible for a somewhat dreary type of country. Ground which was before uneven has often been levelled by the greater accumulation of the clay in the valleys than on the ridges. The summit of the clay when not level often forms low rolling hummocks, and as the clay is usually impervious to water, the ground is often covered by a heavy soil, frequently supporting only rank grass. Accordingly, where boulder-clay occurs in abundance the scenery of the tracts which are occupied by it is somewhat uninteresting.

CHAPTER XXVIII.

THE TERTIARY ROCKS.

4. THE PLEISTOCENE PERIOD.

B. *The Time of Occupation of Britain by Early Man.*

THERE is evidence to shew that at the close of the Glacial period our area stood higher above the sea-level than at present, and it was then connected with the Continent.

During the retreat of the ice no doubt the rivers were greatly swollen, and were capable of much erosion, and of transport of coarse materials. We find accordingly that one of the most important kinds of deposit formed in the times immediately following the Glacial period consisted of river gravels.

Other accumulations, of course, were formed, but the most interesting to us are these gravels, and certain cavern-deposits, for it is in these gravels and the deposits of some caverns that we find the first evidence of the existence of man in Britain. That the gravels which contain the relics of man were formed subsequently to Glacial times is clear for several reasons. Occasionally, though rarely, they are actually found reposing upon Glacial deposits. At other times they rest at various heights in valleys which have been cut through high ground capped with boulder-clay, in such a way as to shew that the boulder-clay rested on the ground before the valley in which the gravels were deposited was formed. In most of the gravels many pebbles are found which can be proved to have been washed out of the boulder-clays, shewing most conclusively that this clay was in existence before the formation of the gravels.

We have had occasion to notice the gradual appearances of higher organisms embedded as fossils as we passed from our study of the older rocks to that of newer beds. In the Cambrian rocks were no animals with backbones. The Silurian rocks have yielded fishes only. Amphibians and reptiles appeared in later strata, and the earliest mammals in still newer deposits; these mammals were lowly forms, and higher mammals appear in the early Tertiary rocks. We can hardly suppose therefore that man had come into existence in early Tertiary times, and no traces of his remains are found in the Tertiary rocks save in some of the most modern deposits. As to when and where man appeared, geology at present gives us no clue; we can only believe from all the evidence that his first appearance on the earth was, when compared with the whole of the time during which the strata were being formed, a very recent event.

No doubt man had been in existence for some time before he reached our area; and also it is possible that the earliest traces of man which have been discovered in Britain are not those of the first men who had reached the country. Be this as it may, the first satisfactory proofs of the existence of man in Britain are found in these Pleistocene deposits of a date posterior to that of the Glacial period. Attempts have been made to prove the occupation of Britain by man in Glacial times, but the proofs are not regarded as satisfactory by those most capable of forming a judgment. It is quite likely, however, that small glaciers existed in the upland valleys of our hill-regions when man first set his foot upon British soil.

Though the appearance of man in Britain is from a geological point of view a modern event, yet as compared with the date to which we assign the first written records of human history there is no doubt that it was very remote. We shall presently give reasons for this conclusion, which are based chiefly on the changes in the physical condition of our area and in the alteration in its living beings which have taken place since man's appearance.

Before discussing these reasons we must say something about the nature of the human relics which have been found in the later Pleistocene deposits, for some years passed after the discovery of the relics before their human origin was admitted.

It is rare to find the actual bones of man in these early deposits, but that is easily explained, for many accumulations have been formed in historic times which are also marked by an absence of human bones. Nevertheless some bones of this early human period have been detected in caverns.

The principal relics consist of various implements fashioned by man, for use as tools and weapons. These relics are made exclusively of stone and bone, and those fashioned of stone are far commoner than are those formed of bone. Not only was the art of making metal tools unknown at this time, but the stone implements were of a far ruder type than those which were fashioned by modern savages in recent times when they were first visited by Europeans, for the modern savages were in all cases acquainted with the methods of rendering their implements more finished by grinding and polishing, whereas the early men of Britain did not even grind or polish the stones which they worked, but merely chipped them into the shape required. Fig. 32 represents a common form of these early British relics, which were in most cases formed of flint, a substance which readily lends itself to being worked into definite shapes.

Fig. 33.

Fig. 32.

Human Implement.
Late Stone Period.

Human Implement.
Early Stone Period.

After glancing at the figure, the reader may well ask why it should be considered that such an object was necessarily fashioned by man, and the answer is not an easy one, for until one has had much experience of the way in which flint breaks as the result both of natural causes and of the intentional act of man, one does not grasp the significance of the nature of the worked implements. Indeed, one who has not had experience will frequently collect flints with surfaces produced by natural processes under the idea that they are the implements of human make, being misled by mere general likeness.

Some of the proofs that these implements are of human make are as follows:—

(i) Study shews that they were formed with the deliberate intention of obtaining definite shapes, and for this purpose flakes were struck from the stone, usually producing more or less parallel surfaces. In the case of the early implements of Britain the shapes produced conform fairly well to three well-marked types.

(ii) These types are discerned among the implements of this age, even in remote areas, the same types as those of Britain being found in France, Spain, Italy, Greece, Palestine, India, and Northern Africa.

(iii) These objects have not been found in deposits in which it is improbable that human relics should be discovered. If they were due to natural causes they should be found in deposits of Primary, Secondary, and early Tertiary age. But they are never found in these rocks, whereas they occur in abundance in many places in the rocks of late Pleistocene date.

(iv) As before mentioned, although actual human bones are rarely associated with these implements, they have been occasionally found.

(v) There is a general similarity between some of

these implements and some which were in use by savages when first discovered by Europeans. This similarity is specially noticeable in comparing the ancient implements with those which were fashioned by the Eskimos.

We will now consider the modes of occurrence of the human relics in the river gravels and in the caverns.

In chapter XI some account was given of the formation of river terraces. It is in some of these terraces that the relics of man are found. Some of the implements have been waterworn since their formation, shewing that they have been rolled along by the river waters, whereas others have their sharp cutting edges preserved, which proves that they were quickly buried in the river gravels. We need not discuss the ways in which they came into the gravel. Anyone who examines the mud at the bottom of an existing river will find human relics which have been embedded in that mud owing to various causes, some being washed into the river, some thrown in, and some again being there owing to the sinking of boats.

Lying often side by side with the human relics are the remains of various animals, which we must briefly notice.

Of those without backbones, the remains of shells of molluscs are frequent. Some of these are land shells, others freshwater shells. They are all forms which are still living, though a few no longer live in Britain.

The remains of animals with bones are most remarkable. They are chiefly those of mammals. Some of these mammals, especially the smaller ones, are still found living in our country, while others have disappeared from Britain, though still living elsewhere; others again are extinct.

Of those living elsewhere we may notice the bones of

the glutton, the reindeer, and the musk ox; which now live in places having a cold climate.

Three most remarkable animals which were living in Britain with early man were the mammoth, the woolly rhinoceros, and the hippopotamus. The mammoth was an elephant, but it is different from the living elephants, and the woolly rhinoceros also differs from the living form of rhinoceros. The remains of the mammoth and woolly rhinoceros have been found frozen in the ice of northern regions with the flesh preserved, and these beasts were protected with thick hairy coverings which enabled them to resist the cold.

The presence of the hippopotamus and of some other beasts which have not been mentioned points to the climate at some times having been warmer than at others during the existence of early man in Britain. This, however, is a topic which is still obscure.

The relics in the caverns differ in some respects from those found in the river-gravels, chiefly in the greater skill displayed in the formation of some of the implements, which otherwise resemble those of the gravels. The same group of mammals is found associated with man's remains in the caverns as in the gravels.

In Britain, human relics have been found, especially in some of the caverns of Derbyshire, and in Kent's cavern near Torquay. Similar relics occur in caverns in France and other countries. Those of France are of great interest, on account of the nature of some of the bone implements. On some of these the workmen have scratched rude representations of some of the beasts which lived at the time. Among these are figures of a large bear, the horse, the reindeer, and the urus (a large ox). The most celebrated implement bears a rude but faithful drawing of the mammoth with its hairy covering actually indicated.

The relics of these early men shew that they belonged to a race who obtained their livelihood by fishing and hunting.

We must now consider more particularly the evidence which points to the remote antiquity of the men who fashioned these implements.

The physical changes which have occurred in our country are very important.

Parts of the river terraces in the Thames valley which have yielded human relics are now fifty feet above the level of the river, and in the neighbourhood of Salisbury implements are found in gravels at a height of a hundred feet above the present river level. From what has been said in chapter XI it is clear that the rivers at one time ran at the levels marked by these terraces, and have eroded their valleys to the present river-levels since the period when early man lived in our area. Not only have these valleys been deepened, but they have in many cases also been widened after the deepening process. Our study of erosion by rivers has led us to the conclusion that the processes of erosion are not rapid, and accordingly the mere fact that the valleys have been enlarged to so great an extent shews that a long period has elapsed since the existence of the men whose relics we have described.

It is true that, as the land stood higher above the sea-level than it does at present, the rivers would have a greater velocity and therefore a greater eroding power than they now possess. The sluggish rivers of our southern counties are at present doing little or nothing in the way of erosion, so we must not judge of their work in past times from observation of what they can or cannot do at the present day. Granting this, however, we cannot suppose that the enlargement of our valleys to the extent mentioned was a rapid process; on the contrary, a long period must have elapsed to allow of the changes.

Again, it has been seen that the assemblage of animals which lived with early man was very different to that now existing in Britain, some of the forms having disappeared from Britain though living elsewhere, while others have become extinct. No doubt man himself assisted in this extinction, but the process cannot have been rapid. So great a change in the assemblage of animals indicates a long period of time for its occurrence. Indeed, we get direct evidence that the extinction was not sudden. Some of the animals disappeared with early man, others lingered until later prehistoric times, and others again were in existence even in the time of the occupation of Britain by the Romans.

The conclusion reached from these and other considerations is confirmed by evidence which is not of a geological character. The study of the progress of civilisation shews that the process of advance from the savage through the barbarous to the civilised stage of society is slow, and therefore we are led once more to believe that a long period has passed since the first occupation of Britain by man.

5. The Recent Period.

Other types of implement besides those briefly described above were formed in Britain before the act of smelting metals had been discovered, and accordingly, the Stone age, which is the term applied to those times when man was unacquainted with the art of making instruments from metal has been divided into two distinct periods so far as Britain and adjoining areas are concerned: these are the Early Stone period which we have just considered, and the Late Stone period to which we are about to call attention.

The implements of the later period shew a far higher degree of finish than those of the earlier one, and are often ground and polished. They also occur in much greater variety, two common types being the arrow-head and one which resembles a broad short chisel, with a cutting edge. An example of the latter type is shewn in Fig. 33.

There is a marked gap between the Early and the Late Stone period in our country. This gap is indicated not only by the abrupt change from the rude to the highly finished types of stone instruments, but also by the great physical changes which had occurred between the early and late periods, for most of those changes to which we briefly alluded as having taken place since the early period were practically completed before the occupation of our area by the men who made the more highly finished types of implements. What this gap indicates is not yet fully explained.

The relics of the Late Stone period are found in valley-bottoms as well as on the intervening ridges, thus shewing that the valleys had been scooped out to their present levels. They sometimes rest upon the surface, at other times are buried in a few feet at most of peat, silt, or soil. At other times again they are found in mounds which were erected as graves for their possessors, and some have actually been found in pits which were dug to extract flint for the fashioning of the implements. Of these the most important occur at Brandon in Suffolk, where the manufacture of gun-flints is still carried on, a survival of the earliest manufacture of our country.

The animals existing in Britain during this period were not unlike those still living. One or two forms have since become extinct, and must have been driven out of our country, but the general assemblage is like that still found.

By degrees the use of metal was introduced from the Continent. First came bronze, and its use practically replaced that of stone, and still later the introduction of iron caused the disuse of bronze.

The first iron implements came into Britain before historic times, and more complex types were introduced or invented by degrees. In the meantime the first historic records enable the historian to take up the story of man's occupation of Britain, and the geologist, who with the student of archæology has discovered the history of man in Britain in early times, now makes way for the historian.

The implements of the later stone, bronze, and iron periods are often associated in the same surface accumulations. Sometimes we find a certain order of succession, the older relics lying below the newer, but sometimes they lie side by side in the rock or on the surface, shewing how little change has occurred in our area since the Later Stone period. Some changes, and those of considerable interest, did take place, but we cannot discuss these here.

It will be noted that no great change in the progress of events from the point of view of the geologist marked the advent of man. When reading the earth's records aright, we find that those slow changes which were taking place in far distant times still went on, and are still in progress. We are living in a geological period, and although this history of the Recent period is the last which the geologist of to-day can write, who can say what periods may follow, marked by change from land to sea, and from sea to land, by outbreak of volcanoes, alteration of rocks, variations of climate, and the coming into existence of assemblage after assemblage of new organisms?

CHAPTER XXIX.

CONCLUSION.

WE have seen that the object of the geologist is to write a history of the earth from earliest times to the present day. As one result of his studies he is able to give information of practical value to those engaged in engineering, mining, quarrying, agriculture, and many other pursuits. But his primary object is nevertheless to restore, as far as possible, the history of the past.

The pursuit is health-giving alike to body and to mind. The memory is quickened as the result of the numerous facts which he must lay up in the storehouse of his mind. The reasoning powers are exercised in a marked degree; in many cases proof of any conclusion at which the geologist arrives is only reached as the result of the acquirement of a considerable mass of evidence, and it requires much cultivation of the reasoning powers before one can say when the evidence which has been gathered is enough to prove the conclusion. The beginner in this science, as in other subjects, must of course take for granted much of what he reads, but the sooner he acquires for himself independence in thought, at first in simple things, the more will geology be useful to him as a mental training.

The powers of observation are also trained by proper study of the science, whether in the museum or in the open country, and in the latter the working of the mind is stimulated by the bodily health acquired by the pleasurable exertion of gathering facts out of doors.

A book like the present one can lay no claim to teach geology, but rather to point out the ways in which the would-be geologist must teach himself. Before one can lay claim to be a geologist, one must go through a long course of training, not only through the medium of books, but also by actual observation. The science can no more be taught by books than can the arts of painting and music, the profession of engineering, or even sports like fishing and hunting. In all of these books are useful, but practice is quite necessary, and so it is in the case of geology.

In the foregoing chapters an attempt has been made to shew in a general way the methods which must be followed for the proper study of the science, by collecting all the facts which can be acquired, and applying our knowledge of these facts in order to try to restore the history of the earth, by studying the rocks of the different periods in their right order. So imperfect is our knowledge of the facts, that even in advanced works only a general idea of the conditions of past times can be given, and in a small work the review of the march of events must of necessity be very incomplete. A hazy idea of the physical history of each period of the past can be acquired as the result of brief study in the case of a limited area, and accordingly we have confined our attention chiefly to the geology of our own country. When we arrive at the Recent period, the present physical conditions are capable of study in much greater detail, and one of the fascinating branches of the science is the attempt to account for existing features after the study of geology.

In order to give a glimpse of this branch we have briefly alluded to the influence of the rocks of various ages as affecting the present scenery of the country, but it would require a separate volume to do justice to this branch of the subject.

I have spoken of the acquirement of a hazy idea of the physical history of each period, as being all that can be done as the result of brief study. The attainment of this, however, is by no means useless. Just as one standing on the summit of some prominent hill in an upland region on a misty day may obtain glimpses of the surrounding hills, vales, rivers and lakes sufficient to inspire him with a desire to learn more of the geography of that country by more detailed exploration, perhaps extending over a period of days, weeks, or even years, so the reader for whom the veil which is thrown over the history of the past is ever so slightly drawn aside may be tempted as the result of further work to learn more of the mysteries which are concealed behind that veil.

To do so he must at once learn to observe for himself. If he has access to a geological museum, well and good. He may be able to compare specimens which he has collected with similar objects preserved and labelled in the museum. But in the case of rocks and fossils much labour must be gone through before he can confidently refer the specimens to their right place, and there is danger in hasty conclusions as to the nature of rocks and fossils. Let the beginner collect specimens by all means, but, if possible, he should seek the help of some friend versed in the science, in order to know what the objects are which he meets with in his expeditions.

But the open country is the true museum for the beginner. Armed with hammer, pocket-lens, note-book, and collecting-bag, let him scour the country in his own neighbourhood, or in any place which he is able to visit.

Then let him study the work of wind, rain, frost, brooks, sea-waves, and, indeed, of any agent which is operating within reach of him: even the runnel on the roadside or on the mud-flats of the shore at low tide will be instructive. Let him also notice the differences of outlines of such hills or ridges as may occur in his district, or if the district be flat, let him consider why it is so. He must notice the wandering of rivers in their curves, the arrangement of shingle, sand, and mud on the sea-coasts, the influence of the underlying rocks upon the soil.

Every natural or artificial exposure of rock should be studied, the nature, and if the rocks are stratified, the order of succession of the rocks, should be noted, and he should seek out why each rock has acquired the particular characters which it possesses.

Everything which he sees which appears to him to throw light upon his studies should be jotted down in his note-book; and sketches as well as notes will assist him in his work.

So, by degrees, the long periods of time with which the geologist deals will be impressed on him in a way which is impossible by a mere statement of the number of million years during which different men of science consider that the earth has been in existence.

When he finds out how slight are the changes which have taken place in his own country as compared with those which have come about since the first appearance of man therein, how thin the accumulations formed since the appearance of man when contrasted with those of Tertiary times, and what a small proportion of the whole of the sedimentary rocks are those of Tertiary age, and lastly, that the earliest of these sedimentary rocks was formed at a time long after the formation of our globe :— he will then get a just idea of the antiquity of the earth.

Happy then will he be, when wandering by upland stream, or beneath high sea-cliff, walking by the shores of the mountain mere or scaling the crags of some high- land hill, to find that each outline of the country, and each rock which crops above the surface of the ground yields to him in whole or in part its fascinating story.

INDEX.

Printed in the United States
By Bookmasters